Biotechnology in Agriculture, Industry and Medicine Series

INDUSTRIAL BIOTECHNOLOGY: PATENTING TRENDS AND INNOVATION

BIOTECHNOLOGY IN AGRICULTURE, INDUSTRY AND MEDICINE SERIES

Agricultural Biotechnology: An Economic Perspective
Margriet F. Caswell, Keith O. Fuglie, and Cassandra A. Klotz
2003. ISBN: 1-59033-624-0

Biotechnology in Agriculture and the Food Industry
G.E. Zaikov (Editor)
2004. ISBN: 1-59454-119-1

Governing Risk in the 21st Century: Lessons from the World of Biotechnology
Peter W.B. Phillips (Editor)
2006. ISBN: 1-59454-818-8

Biotechnology and Industry
G.E. Zaikov (Editor)
2007. ISBN: 1-59454-116-7

Research Progress in Biotechnology
G.E. Zaikov (Editor)
2008. ISBN: 978-1-60456-000-8

Biotechnology and Bioengineering
William G. Flynne (Editor)
2008. ISBN: 978-1-60456-067-1

Biotechnology: Research, Technology and Applications
Felix W. Richter (Editor)
2008. ISBN: 978-1-60456-901-8

Biotechnology, Biodegradation, Water and Foodstuffs
G.E. Zaikov and Larisa Petrivna Krylova (Editors)
2009. ISBN: 978-1-60692-097-8

Industrial Biotechnology
Shara L. Aranoff, Daniel R. Pearson, Deanna Tanner Okun, Irving A. Williamson, Dean A. Pinkert, Robert A. Rogowsky, and Karen Laney-Cummings
2009. **ISBN:** 978-1-60692-256-9

Industrial Biotechnology: Patenting Trends and Innovation
Katherine Linton, Philip Stone, Jeremy Wise, Alexander Bamiagis, Shannon Gaffney, Elizabeth Nesbitt, Matthew Potts, Robert Feinberg, Laura Polly, Sharon Greenfield, Monica Reed, Wanda Tolson, and Karen Laney-Cummings
2009. ISBN: 978-1-60741-032-4

Biotechnology in Agriculture, Industry and Medicine Series

INDUSTRIAL BIOTECHNOLOGY: PATENTING TRENDS AND INNOVATION

KATHERINE LINTON, PHILIP STONE, JEREMY WISE, ALEXANDER BAMIAGIS, SHANNON GAFFNEY, ELIZABETH NESBITT, MATTHEW POTTS, ROBERT FEINBERG, LAURA POLLY, SHARON GREENFIELD, MONICA REED, WANDA TOLSON AND KAREN LANEY-CUMMINGS

Nova Science Publishers, Inc.
New York

Copyright © 2009 by Nova Science Publishers, Inc.

All rights reserved. No part of this book may be reproduced, stored in a retrieval system or transmitted in any form or by any means: electronic, electrostatic, magnetic, tape, mechanical photocopying, recording or otherwise without the written permission of the Publisher.

For permission to use material from this book please contact us:
Telephone 631-231-7269; Fax 631-231-8175
Web Site: http://www.novapublishers.com

NOTICE TO THE READER

The Publisher has taken reasonable care in the preparation of this book, but makes no expressed or implied warranty of any kind and assumes no responsibility for any errors or omissions. No liability is assumed for incidental or consequential damages in connection with or arising out of information contained in this book. The Publisher shall not be liable for any special, consequential, or exemplary damages resulting, in whole or in part, from the readers' use of, or reliance upon, this material.

Independent verification should be sought for any data, advice or recommendations contained in this book. In addition, no responsibility is assumed by the publisher for any injury and/or damage to persons or property arising from any methods, products, instructions, ideas or otherwise contained in this publication.

This publication is designed to provide accurate and authoritative information with regard to the subject matter covered herein. It is sold with the clear understanding that the Publisher is not engaged in rendering legal or any other professional services. If legal or any other expert assistance is required, the services of a competent person should be sought. FROM A DECLARATION OF PARTICIPANTS JOINTLY ADOPTED BY A COMMITTEE OF THE AMERICAN BAR ASSOCIATION AND A COMMITTEE OF PUBLISHERS.

LIBRARY OF CONGRESS CATALOGING-IN-PUBLICATION DATA
Available upon Request.

ISBN: 978-1-60741-032-4

Published by Nova Science Publishers, Inc. ✝ New York

CONTENTS

Abstract		vii
Executive Summary		1
Chapter 1	Introduction	7
Chapter 2	Patents and Innovation	13
Chapter 3	Aggregate Data and Questionnaire Results	21
Chapter 4	Industrial Biotechnology at the Technology and Firm Level	33
Chapter 5	Conclusions	61
Appendix A.	Patent Search Methodology	73
Index		77

ABSTRACT*

This study provides a profile of innovation in industrial biotechnology, an emerging field of biotechnology characterized by the use of enzymes, microorganisms, and other biocatalysts to create new processes and products. This study uses patent data, survey results, and technology and firm level data from emerging sectors of industrial biotechnology to provide a detailed picture of innovation in the field.

This study finds substantial evidence that the field of industrial biotechnology is diverse and growing, with new patent owners entering at a steady rate. Different companies, ranging from small to large in size, are dominant in different areas of industrial biotechnology and patent portfolios play an important role in their participation by facilitating the commercialization of new products and processes. Moreover, according to the U.S. International Trade Commission's recent report, most firms in the biofuel and chemical industries state that patent barriers are not creating a substantial impediment to the research and development (R&D) or commercialization of industrial biotechnology products.

The U.S. Patent and Trademark Office (USPTO) issued 20,428 utility patents with a primary classification related to industrial biotechnology from January 1975 through December 2006. The number of patents issued each year climbed steadily beginning in the mid-1980s, peaked in 1999, declined

---

* This is an edited, excerpted and augmented edition of a Office of Industries, U.S. Internation Trade Commision publication, Staff Research Study. Publication 4039, dated October 2008.

from 2000 through 2005, and rebounded in 2006. The trends for industrial biotechnology patenting mirror those in the broader field of biotechnology and are strongly influenced by capacity and resource issues at the USPTO, particularly longer and more rigorous review periods.

The study's focus on two emerging areas of industrial biotechnology—cellulaserelated enzymes used in the production of cellulosic ethanol and enzymes for biobased plastics production—enables a closer look at the role patents are playing at the technology and firm levels. In both technology areas, leading patenting firms hold a relatively small number of patents. The Herfindahl-Hirschman Index indicates that the control of patents rights in these emerging fields is not concentrated; a steady stream of new owners has been entering the fields over the last ten years.

Profiles of leading patenting firms in these emerging fields—Novozymes, Verenium, Metabolix, and Cargill—illustrate the important role that patents are playing in firm activities. They protect R&D investments and market share, and facilitate the strategic alliances with other firms and the federal government that provide the know-how and capital to move cellulosic ethanol and bio-based plastics technologies from R&D to commercialization.

EXECUTIVE SUMMARY

This study provides a profile of innovation in industrial biotechnology, an emerging field of biotechnology characterized by the use of enzymes, microorganisms, and other biocatalysts to create new products. Industrial biotechnology is used to make biofuels, chemicals, and other products in more sustainable and environmentally friendly ways by, for example, enabling the use of renewable resources rather than petroleum-based products, eliminating harmful byproducts created by conventional chemical processes, reducing energy requirements and greenhouse gas emissions, and/or lowering manufacturing costs. Because of these positive attributes, the demand for industrial biotechnology products and processes is increasing. This study uses patent data, survey results, and technology and firm level data from emerging sectors of industrial biotechnology to provide a detailed picture of innovation in the field.

Although patents have been shown to facilitate the movement of new products and processes from research and development (R&D) to commercialization, the literature raises an important question: is there evidence that too much patenting stifles innovation? This could occur, for example, when firms amass huge patent portfolios or when there are a very large number of patents in a narrow technology area, making it difficult for companies entering the market to find patentable space. The authors found that innovation is expanding in the emerging field of industrial biotechnology. A diverse group of firms, large and small, is developing new patented products and processes, and new firms are steadily entering the field. Moreover, the average number of patents held by firms active in the field is still relatively small. These data suggest that patents are facilitating and not stifling innovation.

This study profiles four firms that are leading patent owners in two emerging sectors of industrial biotechnology: cellulase-related enzymes used to break down cellulosic biomass to produce biofuels and chemicals, and the production of bio-based plastics using enzymatic processes. The profiles describe the firms' operations, R&D efforts, and their strategic alliances with other firms and government, highlighting the important functions that patents play in these activities.

KEY FINDINGS: AGGREGATE PATENT DATA AND QUESTIONNAIRE RESULTS

This study relies on a variety of data sources to identify the role of patents in industrial biotechnology innovation. Chapter 3 reviews aggregate data obtained from a search of U.S. Patent and Trademark Office (USPTO) patent records for the period January 1975 through December 2006, and results of a questionnaire sent to firms in the biofuel and chemical industries. Key findings are set forth below.

- The USPTO issued 20,428 utility patents with a primary classification related to industrial biotechnology from January 1975 through December 2006. The number of patents issued each year climbed steadily beginning in the mid-1980s, peaked in 1999, declined from 2000 through 2005, and rebounded in 2006. The trends for industrial biotechnology patenting mirror those in the broader field of biotechnology and are strongly influenced by capacity and resource issues at the USPTO, particularly longer and more rigorous review periods.

- Patents obtained by domestic universities grew substantially over the period, coinciding with implementation of the Bayh-Dole Act. This act permits universities to retain patent ownership in government-funded inventions and license those inventions to others. It has resulted in a substantial upswing in university patenting.
- Patents in industrial biotechnology do not appear to be concentrated in the hands of a small number of owners and new owners are steadily entering the field. Even the number of patents held by industry leaders is relatively small, particularly when compared with the much larger patent portfolios of leading patenting companies in information technology and other high tech areas.
- More than 70 percent of biofuel and chemical company representatives responding to a survey by the U.S. International Trade Commission (Commission) reported that "patent barriers" are one of the least significant impediments to the R&D and commercialization of industrial biotechnology products and processes.

KEY FINDINGS: INDUSTRIAL BIOTECHNOLOGY AT THE TECHNOLOGY AND FIRM LEVEL

This study next focuses on two emerging areas of industrial biotechnology— cellulase-related enzymes and enzymes for bio-based plastics production—to enable a closer look at the role patents are playing at the technology level (as determined by particular patent classifications) and at firm levels. Key findings from chapter 4 are presented below.

Cellulase enzymes are critical to achieving the cost-effective production of cellulosic ethanol. A large portion of the 277 cellulase-related patents issued by the USPTO in classification 43 5/209 is held by two Danish companies, Novo Nordisk A/S (a related company to Novozymes) and Danisco A/S (the parent company of Genencor). While these companies are dominant in this class, a number of smaller companies also own multiple patents. Moreover, there is a steady stream of new owners entering the field.

Application of the Herfindahl-Hirschman Index (HHI) indicates control of patent rights in the field is not concentrated. The existence of many patenting firms, with most having only one patent, lowers the index and keeps the market from being deemed concentrated.

- The Novozymes profile highlights the important role that a strong patent portfolio plays in the firm's activities. According to Novozymes, patents protect its R&D investments and market share, and secure the company's operational freedom. Novozymes also has recently prevailed in high- stakes patent litigation with its chief competitor, Genencor, over enzymes used in the production of biofuels.
- Verenium, a small firm that owns numerous patents for enzyme technologies and has licensed foundational cellulase-related patents from the University of Florida, is also profiled. Patents facilitated the transfer of the university technology to Verenium for further development. They also have helped Verenium enter into strategic alliances with firms and the federal government, which have provided the capital needed to move its cellulosic ethanol technologies from R&D to commercialization.
- Bio-based plastics, particularly poly-hydroxyalkanoates (PHA) and polylactides (PLA), is another emerging sector of industrial biotechnology where patents play a central role in innovations. The USPTO issued 308 patents in 435/135, the classification related to PHA. Most companies patenting in this class hold a small number of patents and the HHI indicates that the control of patent rights in the field is unconcentrated.
- The firm with one of the largest patent holdings in the PHA-related classification, Metabolix, is a small U.S. firm. The profile of Metabolix highlights a strong patent portfolio, based on foundational technology acquired from the Massachusetts Institute of Technology (MIT). For Metabolix, patents have facilitated both the transfer of early-stage university research to the company and the strategic

alliances and government grants that bring necessary capital for the commercialization of bio-based plastics.
- The USPTO issued 69 patents in 435/139, the classification related to PLA, another bio-based plastic. Most corporations hold only 1 patent in this classification and even the top patenting companies hold less than 10 patents. The largest patent holder is a privately held U.S. firm, Cargill. The profile of Cargill shows that its PLA-related patents, and those of its subsidiary NatureWorks, support a strong market position. NatureWorks operates the only world-scale PLA plant in the United States.

Together the aggregate data and firm questionnaire results in chapter 3 and the technology and firm level data for particular sectors described in chapter 4 highlight the central role that patents play in innovation in industrial biotechnology. The firms profiled rely on patents to enhance and protect their market share, transfer technologies from the university setting, attract technology and outside funding from strategic alliance partners and the federal government, and protect substantial R&D investments. A diverse and expanding group of companies, large and small, are relying on patent portfolios and strategic alliances to move industrial biotechnology products and processes from R&D to commercialization.

Chapter 1

INTRODUCTION

This study provides an in-depth look at patenting in the field of industrial biotechnology, building on the Commission's recent report on the development and adoption of industrial biotechnology by the U.S. chemical and biofuel industries.[1] Patenting in this sector of biotechnology has received little attention; this study seeks to fill that gap.

This study fits within a rich literature on the relationship between patents and innovation, including the particularly important role that patents have played in the development of the field of biotechnology. An important question suggested by the literature, and to which this study is responsive, is whether there is evidence that patents are stifling innovation by keeping new entrants out of the field. This study finds substantial evidence that the field of industrial biotechnology is diverse and growing, with new patent owners entering at a steady rate. Different companies, ranging from small to large in size, are dominant in different areas of industrial biotechnology, and patent portfolios appear to play an important role in their participation. Moreover, as noted in the Commission's recent report, most firms in the biofuel and chemical industries state that patent barriers are not creating a substantial impediment to the R&D or commercialization of industrial biotechnology products.

This study presents patent data obtained from a custom data set provided by the USPTO based on search criteria developed by the authors. The data are used as a basis for discussing patenting trends and the characteristics of patent owners. Also presented are the results of the Commission's questionnaire of firms in the biofuel and chemical industries, and particularly their responses to questions about whether patent barriers are a significant impediment to the R&D and commercialization of industrial biotechnology products.

This study also describes patenting trends and ownership characteristics in two emerging areas of industrial biotechnology in the biofuel and chemical industries: the creation and use of enzymes that break down cellulosic biomass for the production of chemicals and biofuels, including cellulosic ethanol, and the production of bio-based plastics using enzymatic processes. Profiles of four of the leading patenting companies in these fields are provided; these profiles highlight the important role that patents play in the movement of new products and processes from R&D to commercialization. This study paints a new and detailed picture of patenting and innovation in industrial biotechnology in the biofuel and chemical industries.

SCOPE: BIOTECHNOLOGY AND INDUSTRIAL BIOTECHNOLOGY

Biotechnology is a collection of technologies that rely on the use of cellular and biomolecular processes to develop or make useful products.[2] Industrial biotechnology, or white biotechnology, is called biotechnology's "third wave," following more well-known applications in the medical sector (red biotechnology) and agriculture (green biotechnology).[3] Although there are many definitions of industrial biotechnology, for the purposes of this study and as defined in the Commission's report, the touchstone is the use of enzymes or micro-organisms to catalyze chemical reactions.[4] These enzymes or micro-organisms (often referred to as "biocatalysts") typically are used to manufacture intermediate and end products more efficiently, reduce environmental impacts of processes and products, and/or enable the creation of new products from renewable resources.[5]

The use of enzymes to catalyze chemical reactions is not new; naturally occurring enzymes have been used for centuries in the production of cheese and other foods. Crude enzyme extracts were used in the industrial processing of textiles and leather in the late 19th and early 20th centuries.[6] Within the past 50 years, increasingly sophisticated and functional enzymes have been created and integrated into chemical processes and products by the chemical industry. By the 1960s, for example, commercially successful enzyme-based detergents and cleaners had been developed.[7] The 1 960s also saw the development of enzymes to break down starch from corn or other crops into its component sugars. These sugars would then be converted to products, such as ethanol or sweeteners. Enzymatic processes typically replaced the acid hydrolysis of starch and provided higher yields and greater purity.[8] The latest versions of these enzymes are used today in the production of corn-based ethanol.

Advances in the biological sciences in the 1980s and 1990s allowed for the genetic manipulation of micro-organisms to produce biocatalysts with more targeted and specialized functions. Within the chemical industry, such specialized biocatalysts are being used to produce an increasingly diverse set of chemicals, ranging from pharmaceuticals, including vitamin B^{12} and cephalosporins, to bio-based plastics. In the biofuels industry, specialized biocatalysts are enabling the production of cellulosic ethanol and other biofuels and helping make production more cost effective.[9]

Today, industrial biotechnology companies like the ones profiled in this study—Novozymes, Verenium, Metabolix, and Cargill—typically look for natural biocatalysts; identify and improve them for specific applications using screening, genetic engineering, and other high technology processes; and manufacture them in commercial quantities using fermentation.[10] Companies active in many industry sectors, including biofuels, fine and bulk chemicals, pharmaceuticals, food and beverages, detergents and household products, and textiles use these enzymes to improve the environmental, technical, and economic performance of their existing products and processes, and to create new products and processes.[11]

DATA

The data for this study were drawn from a variety of sources. The first source is the custom data set provided by the USPTO.[12] The data set identifies 20,418 patents granted by the USPTO during the 1975–2006 period in subject matter classifications identified by the authors as related to industrial biotechnology in the biofuel and chemical industries. A second source for patent data was the USPTO Patent Full-Text and Image Database (USPTO online database), which was used to conduct classification- and company-based searches of patent grants and applications.[13] Unlike the custom data set, the USPTO online database permitted retrieval of information about published patent applications. Also presented are relevant results from the Commission's questionnaire of firms in the biofuel and chemical industries. Firm level data, including information about firms' finances, R&D focus areas, and patent portfolios, were obtained from publicly available sources and from the proprietary Orbis Companies Database.

Because this study relies primarily on patent data, it is important to recognize the inherent advantages and limitations of this type of data. Patent data are readily available through the USPTO and other patent offices, and contain substantial detail about firms' activities; additionally, these data are historical, permitting the analysis of trends. Data limitations include the fact that not all inventions are patented, many patents are never commercialized or licensed, patent applications and granted patents substantially lag underlying inventions, and patent applications can overstate levels of successful innovation because a patent ultimately may not be issued.[14] Bearing in mind these limitations, patent data can provide a useful window into innovative activities.

Data interpretation issues also arise from the substantial overlap between industrial biotechnology and other types of biotechnology. For example, genetically modified crops (green biotechnology) often are used to produce biofuels through enzymatic processes (white biotechnology), and both enzymatic and genetic processes are used in the production of pharmaceuticals (white and red biotechnology). Distinctions made by the biotechnology industry between different types of biotechnology do not fit neatly into the USPTO classification system.

To identify classifications particularly relevant to industrial biotechnology in the biofuel and chemicals industries, the authors developed a list of "model patents" reflective of important discoveries identified in the Commission's recent report.[15] Screening of the classifications associated with these model patents revealed that discoveries in industrial biotechnology and in other types of biotechnology often fell within the same classifications. For example, one of the most common classifications in the model patents was 435/69.1: "Recombinant DNA technique included in method of making a protein or polypeptide," a classification that is also closely aligned with genetic engineering and red biotechnology. This study's additional analysis of particular classifications more closely tied to industrial biotechnology, i.e., those for cellulase- and bio-based plastics-related enzymes, and profiles of top patenting firms in these fields, is intended to bring industrial biotechnology into sharper focus.

REPORT ORGANIZATION

This report is divided into four chapters that together address patenting in industrial biotechnology. Chapter 2 provides a brief overview of the literature on patents and innovation, with a particular focus on the role of patents in the development and growth of the biotechnology industry and the question of whether too much patenting is stifling innovation. Chapter 3 presents the aggregate data obtained from the USPTO with a discussion of trends and characteristics of companies patenting in industrial biotechnology. It also includes the results of the Commission's questionnaire regarding patent barriers and other impediments to R&D and commercialization. Chapter 4 focuses on particular emerging sectors in industrial biotechnology to provide a clearer picture of patenting in these fields and how patent portfolios impact the activities of leading companies profiled. Chapter 5 brings together conclusions arising from the analysis of the literature, aggregate data, questionnaire results, and technology and firm level information, and provides suggestions for future research.

ENDNOTES

[1] USITC, Industrial Biotechnology: Development and Adoption by the U.S. Chemical and Biofuel Industries, USITC Publication 4020, 2008.

[2] BIO, "Biotechnology: A Collection of Technologies," 2008.

[3] USITC, Industrial Biotechnology, 2008, 1–3.

[4] Enzymes are organic compounds that initiate or accelerate chemical reactions. Microorganisms are simple life forms that use enzymes to consume raw materials as part of their metabolism. Examples of microorganisms include bacteria (e.g., E. coli) and yeast. Ibid., 1- 4–1-5.

[5] BIO, "Industrial and Environmental Applications," 2008.

[6] Kirk, et al., "Enzyme Applications, Industrial," 2004, 251.

[7] Ibid., 252.

[8] Ibid.

[9] BIO, "Industrial and Environmental Applications," 2008.

[10] Ibid.

[11] Ibid.

[12] USPTO, Utility Patents, 2007.

[13] USPTO, USPTO Patent Full-Text and Image Database.

[14] NAS, Industrial Research and Innovation Indicators, 1997, 24; NSB, Science and Engineering Indicators 2008, 2008, 6–38.

[15] USITC, Industrial Biotechnology, 2008. Because this research had its inception in the Commission's study of the impact of industrial biotechnology on two of the primary industries in which it is used, biofuels and chemicals, the findings likely overemphasize these applications. The USPTO classifications relied upon, however, are science- or technology- based and not industry-based; thus, the results include applications of the technologies across industries. See app. A for discussion of the search methodology.

Chapter 2

PATENTS AND INNOVATION

A patent is an agreement between the owner of an invention and a country that permits the owner to exclude others from making, using, selling, or importing the claimed invention for a period of time (usually 20 years) in exchange for public disclosure of the invention.[1] The economic view of patents is that they encourage innovation by offering a trade-off; in return for a period of market exclusivity, the inventor must disclose the details of the invention so that others can build on the knowledge disclosed.[2]

There is a rich literature, albeit reaching mixed conclusions, on how well the relationship between patents and innovation works in practice.[3] This literature suggests that, on the one hand, patents assist in the movement of a product or process from the research stage to commercialization by facilitating technology transfer and financing. On the other, it suggests that too many patents may thwart innovation particularly in industries (and biotechnology may be an example) where the innovation process is sequential and cumulative.[4] The advantages and disadvantages of patent protection in biotechnology have been addressed in the literature, with particular attention given to the impact of milestones in biotechnology patent policy.[5] This chapter briefly reviews this literature.

Milestones in Biotechnology Patent Policy

The Supreme Court's 1980 decision in *Diamond v. Chakrabarty*, 447 U.S. 303, 309 (1980), which overturned a ban on the patenting of life forms to uphold a patent on a genetically engineered, petroleum-eating bacterium, is considered a milestone in the development of the biotechnology industry. According to industry representatives interviewed by the Federal Trade Commission (FTC) for a study of competition and patent law policy, the biotechnology industry would not have emerged "but for the existence of predictable patents," and the Supreme Court's decision spurred significant industry growth.[6]

Passage of the Bayh-Dole Act in 1980[7] also is believed to have spurred biotechnology patenting and innovative activity. The act permitted universities and small firms to retain title to inventions financed by the federal government. Prior to this, patent rights often stayed with the federal agency funding the research, and many inventions were not commercialized. In 1983, Bayh-Dole rights were extended to all government contractors as well. In 1984, Congress expanded the rights of universities further by removing restrictions in Bayh-Dole on the kinds of inventions they could own and on their rights to license those inventions to others.[8]

According to Henderson, Jaffe, and Trajtenberg (1998), university patenting "exploded" after the enactment of Bayh-Dole.[9] While the authors considered it difficult to assign roles of cause and effect because university patenting was on the rise even before the enactment, they concluded that "continued exponential growth probably could not have been sustained without removal of cumbersome barriers to patents from federal research."[10] The licensing of university technologies to firms and the creation of spin-off firms from university research, particularly in biotechnology, also grew substantially after the enactment of Bayh-Dole (box 2.1).

Box 2.1. Bayh-Dole and the increase in patenting, licenses, and the formation of spin-off companies

- In 1980, the USPTO issued only 390 patents in all technology areas to U.S. universities. In 2005, universities obtained 2,725 patents, an increase of approximately 600 percent during a time when total patents issued to corporate U.S. inventors increased by approximately 150 percent. Many university patents have been issued in biotechnology-related classes including class 435, which is the most common class for industrial biotechnology patents.
- University licensing of patented technologies also has increased sharply. In 1991, the first year of licensing survey data, members of the Association for University Technology Managers (AUTM) reported entering into 1,278 new licenses for university-developed technologies. By 2006, that number had grown to 4,963 new licenses entered into that year.
- More than 5,000 spin-off companies (many of which have been in the field of biotechnology) have formed around university research since Bayh-Dole's enactment.

Sources: NSB, Science and Engineering Indicators 2008, 2008, 5-40 and 5-41; NSB, Science and Engineering Indicators 1993, 1993, 5-27 and 6-12; AUTM.

At about the same time that Bayh-Dole was enacted, Congress also passed the Stevenson-Wydler Technology Innovation Act of 1980,[11] making technology transfer a priority at the federal laboratories.[12] This act was followed by the Federal Technology Transfer Act in 1986,[13] permitting government research facilities to enter into cooperative R&D arrangements with industry and grant industry title to any resulting inventions. Patenting activities and licensing of technologies developed at federal laboratories grew rapidly in the wake of this legislation.[14]

The impact of these judicial and legislative changes is considered to have been particularly strong in the field of biotechnology. Key genetic engineering techniques and other enabling technologies often were developed with federal funds at universities and research laboratories. These foundational technologies then were acquired by start-up firms and other new entrants to the field for further commercial development.[15]

Biotechnology Patents and the Commercialization of Products

Patents facilitate the transfer of knowledge and technologies from the research setting to commercial development. For example, Jensen and Thursby have demonstrated, based on a survey of university technology managers, that university inventions often are early stage efforts—little more than a proof of concept or prototype.[16] At this stage, technology managers stated that commercialization would only be possible if the inventor were able to license the technology to a commercial developer who could move the invention forward. Jensen and Thursby found that, absent the opportunity to patent and license university technologies, much federally funded research would never be transferred to industry.[17]

The use of patents to facilitate the transfer of technology is particularly important given the increasingly collaborative nature of innovation. Based on a sample of innovations recognized by R&D Magazine as being among the top 100 innovations of the year over the last 40 years, Block and Keller found that approximately 80 percent of award-winning innovations in the 1970s came from large firms acting on their own.[18] By contrast, today approximately two-thirds come from collaborations between firms, universities, federal laboratories, and government agencies.[19] Patents, and other types of intellectual property, facilitate these increasingly frequent collaborations by providing the foundation for the transfer of technology and knowledge between firm, university, and government actors.

The literature further shows that patents facilitate investment in the biotechnology industry and provide a benchmark for potential investors to use in evaluating a firm's technology and likelihood of commercial success. Lerner examined financing rounds for start-up biotechnology firms and found that the market valuation of the firms reflected the scope of their patents.[20]

Patent scope, measured by the number of classification groups to which a patent was assigned, reportedly had an economically and statistically significant impact on the valuation of start-up biotechnology firms in financing rounds. While many factors other than intellectual property affect the valuation of biotechnology firms, particularly as a firm's product approaches the marketplace, Lerner found that intellectual property is the most valuable asset of the young biotechnology company.[21]

Biotechnology Patents and the Anticommons Theory

Not all assessments of the impact of biotechnology patents on innovation are positive. There is some research to suggest that biotechnology, and biomedical innovation in particular, are susceptible to what Heller and Eisenberg have labeled a "tragedy of the anticommons." This situation arises when too many people own property rights in separate inputs needed for the development of a line of research or a product.[22] Shapiro similarly posited that the patent system is creating a dense web of overlapping rights that a company must "hack its way through" to commercialize technology; "stronger patent rights can have the perverse effect of stifling not encouraging innovation."[23] Heller and Eisenberg applied these theories to biotechnology and asserted that a proliferation of rights in upstream technologies, such as gene fragments or the receptors used to screen potential pharmaceutical products, blocks downstream activities because of the need to negotiate and obtain multiple licenses from numerous inventors.[24]

The National Academies of Science (NAS) commissioned a series of studies to consider the question of the effects of patenting on biomedical research and, more particularly, whether there is an anticommons problem. In the first study, Walsh, Arora, and Cohen interviewed professionals in the biomedical field, including attorneys, scientists, business managers, and university researchers, and found little evidence that research had been impeded.[25] An important exception was in the area of genetic diagnostics, where some respondents indicated that patents were interfering with research. However, most survey respondents reported no cases in which valuable projects were suspended because of patent restrictions. Instead, firms and universities reported that they had developed "working solutions" to enable research in heavily patented space, including licensing, inventing around patents, invoking research exemptions to infringement, developing and using public tools, and challenging patents in court.[26]

In a follow-on survey, Walsh, Cho, and Cohen questioned more biomedical professionals and confirmed the earlier finding that only a small number of projects (about 1 percent) were being abandoned or delayed because of difficulties in accessing technology.[27] Thus, the anticommons problem did not appear to be significant. They cautioned, however, that the patent landscape was likely to become considerably more complex over time and that researchers who were proceeding despite concerns about the possibility of patent infringement were likely to become more cautious if the legal environment became more threatening.[28]

More recently, Adelman and DeAngelis conducted an empirical study of biotechnology patents also intended to test the anticommons hypothesis.[29] They identified a set of more than 52,000 biotechnology patents issued during the January 1990–December 2004 period. They found little evidence to support an anticommons effect in biotechnology patents. Ownership of the patents in their database was diverse, with even large companies granted a relatively small number of patents per year, and the number of new entities obtaining patents increasing steadily over the time period.[30] They posited that, unlike the traditional commons scenario, the commons for biotechnology is not finite but relatively unbounded. "Biotechnology methods have produced vast quantities of genetic data, but scientists have not been able to keep up with the explosion of new information"; under these circumstances they found a substantial reduction in the potential for anticommons problems.[31]

Adelman and DeAngelis also emphasized the complexities inherent in the interpretation of patent data and the resulting importance of bringing together various types of information, including qualitative and firm level information about particular advances in the field, to obtain a more accurate picture.[32] Following the trail suggested by Adelman and DeAngelis, this study applies a multifaceted approach to the consideration of patenting in industrial biotechnology. Aggregate data obtained from a search of patents in classifications related to industrial biotechnology are presented, as well as the results of a Commission questionnaire sent to firms in the biofuel and chemical industries. Moreover, patent data and qualitative information at the technology and firm level show the important roles that patents are playing in moving industrial biotechnology products and processes from R&D to commercialization.

ENDNOTES

[1] USPTO, *What are Patents, Trademarks, Servicemarks, and Copyrights?* May 12, 2004.
[2] Fromer, "Patent Disclosure," 2009 (forthcoming).
[3] For example, see Hahn, "An Overview of the Economics of Intellectual Property Protection," 2005, 11–44.
[4] For example, see Heller and Eisenberg, "Can Patents Deter Innovation?" May 1, 1998, 698; Shapiro, "Navigating the Patent Thicket," 2001, 120; and Jaffe and Lerner, *Innovation and its Discontents*, 2004, 1–25.
[5] For example, see Barfield and Calfee, *Biotechnology and the Patent System*, 2007.
[6] FTC, *To Promote Innovation*, 2003, 3–17.
[7] Patent and Trademark Act Amendments of 1980, Pub.L.No. 96-517, 94 Stat. 3014 (1980).
[8] Schacht, R&D Partnerships and Intellectual Property, 2000.
[9] Henderson, Jaffe, and Trajtenberg, "Universities as a Source of Commercial Technology," February 1998, 119.
[10] Ibid., 122.
[11] Pub. L. No. 96-480, 94 Stat. 2311, 15 U.S.C. 3701 (1980).
[12] Jaffe and Lerner, "Reinventing Public R&D," 2001, 170.
[13] Pub. L. No. 99-502, 100 Stat. 1785 (1986).
[14] Jaffe and Lerner, "Reinventing Public R&D," 2001, 194.
[15] Lerner, "Small Businesses, Innovation, and Public Policy," 1999, 159–68.
[16] Jensen and Thursby, "Proofs and Prototypes for Sale," March 2001, 240–41.

[17] Ibid., 255.
[18] Block and Keller, Where Do Innovations Come From? July 2008, 2–3.
[19] Ibid., 3.
[20] Lerner, "The Importance of Patent Scope," 1994, 319.
[21] Ibid., 325.
[22] Heller and Eisenberg, "Can Patents Deter Innovation?" May 1, 1998, 698. The theory is a mirror of the "tragedy of the commons" theory, which holds that if people hold property in common, with no person having the right to exclude others, the property will tend to be overused because there is no incentive to conserve. Conversely, the tragedy of the anticommons posits that when too many people own a resource it is underused because of the transaction costs inherent in reaching agreement with all of the owners.
[23] Shapiro, "Navigating the Patent Thicket," 2001, 120.
[24] Heller and Eisenberg, "Can Patents Deter Innovation?" May 1, 1998, 698.
[25] Walsh, Arora, and Cohen, "Effects of Research Tools Patents and Licensing on Biomedical Innovation," 2003, 285–86.
[26] Ibid.
[27] Walsh, Cho, and Cohen, "View from the Bench," September 23, 2005, 2002–03.
[28] Ibid.
[29] Adelman and DeAngelis. "Patent Metrics," 2007, 1677–1744.
[30] Ibid., 1694–95.
[31] Ibid., 1699.
[32] Ibid., 1727.

Chapter 3

AGGREGATE DATA AND QUESTIONNAIRE RESULTS

This chapter presents patent data and questionnaire results from firms active in industrial biotechnology. The patent data were generated by a search conducted by the USPTO, based on parameters set by the authors, of all patents granted during the January 1975–December 2006 period. The chapter begins with a description of the search parameters and then presents the results, focusing on trends and the ownership characteristics of those patenting in industrial biotechnology-related fields.[1] The chapter next describes results of a questionnaire of biofuel and chemical companies carried out by the Commission in the fall of 2007. The questionnaire included questions about whether patent barriers were impediments to firms' R&D and commercialization of industrial biotechnology products and processes.

PATENT SEARCH PARAMETERS

The search of the USPTO database was classification based, relying on a set of classifications in the U.S. Patent Classification System (USPC) that the authors found to be associated with model patents in different areas of industrial biotechnology.[2] The USPC contains classes and subclasses; classes generally delineate one technology from another, and subclasses (of which there are more than 150,000) delineate processes and features of the subject matter encompassed by the class. Each patent is assigned a single, primary classification indicative of the main inventive concept; patent information that is separately classifiable apart from the primary classification may receive a secondary classification.[3] The USPTO search used in this study was limited to those patents that had a primary classification identified in our search terms.

Based on these model patents from different areas of industrial biotechnology, the following USPC classes were identified as containing potentially relevant subclasses: 435: "Chemistry: molecular biology and microbiology" (most subclasses were within this class); 536: "Organic compounds"; and 127: "Sugar, starch, and carbohydrates." Most of the classifications relied upon in the search fell under subclass 435/41, which covers the production of a desired chemical using micro-organisms, tissue cultures, or enzymes; and subclass 435/183, which covers the production of particular enzymes and micro-organisms.[4] Emerging technologies in industrial biotechnology, including the creation and use of enzymes used in the production of cellulosic ethanol and those related to biobased plastics, fall within these subclasses. The identified subclasses cover much of the activity in industrial biotechnology, as well as some pharmaceutical and agricultural biotechnology activity.[5]

PATENT SEARCH RESULTS

Trends in Patent Grants

During the January 1975–December 2006 period, the USPTO granted 20,418 invention patents with a primary classification relevant to industrial biotechnology. Figure 3.1 shows that the number of patents granted each year climbed steadily in the second half of the 1980s and throughout the 1990s, peaking at 1,688 patents granted in 1999. Thereafter, grants generally declined until they rebounded in 2006, but they did not reach the levels of the late 1990s.

The trends for industrial biotechnology patents mirror those for all biotechnology patenting, according to studies of all biotechnology patents conducted by the USPTO (figure 3.2) and Adelman and DeAngelis.[6] The USPTO reported 112,360 patents granted in all biotechnology fields for the January 1975– December 2005 period.[7] Industrial biotechnology patents, as identified in this study comprised, on average, 17 percent of all biotechnology patents during this period.

There are multiple explanations for the trends identified in figures 3.1 and 3.2. The peak in patenting in the late 1990s has been attributed primarily to a rush in filings prior to the June 1995 change in the U.S. patent term from 17 years from the date of patent issuance to 20 years from the date of application, as required by the Uruguay Round of the General Agreement on Tariffs and Trade.[8] Biotechnology inventors reportedly perceived the 17-year term to be more advantageous because it ran from the date of patent issuance. The pending period could thus be used to comply with substantial product development and regulatory requirements without reducing the patent term, as would occur once the term ran from the date of application. The 1995 peak in applications is reflected in a subsequent peak in patent grants in the late 1990s because the review process for biotechnology patents took about three years during this time.[9]

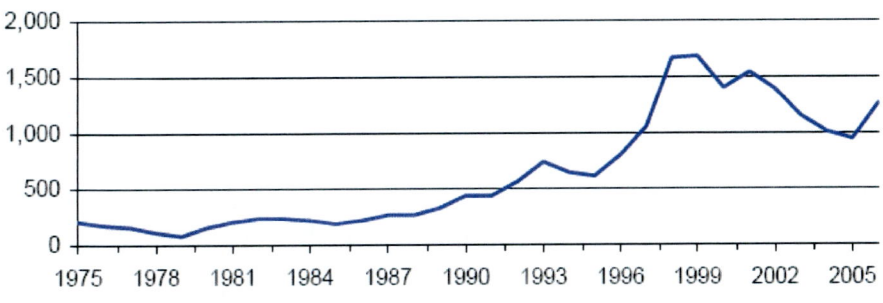

Source: USPTO, Utility Patents, 2007.

Figure 3.1. Industrial biotechnology patents granted yearly, 1975–2006.

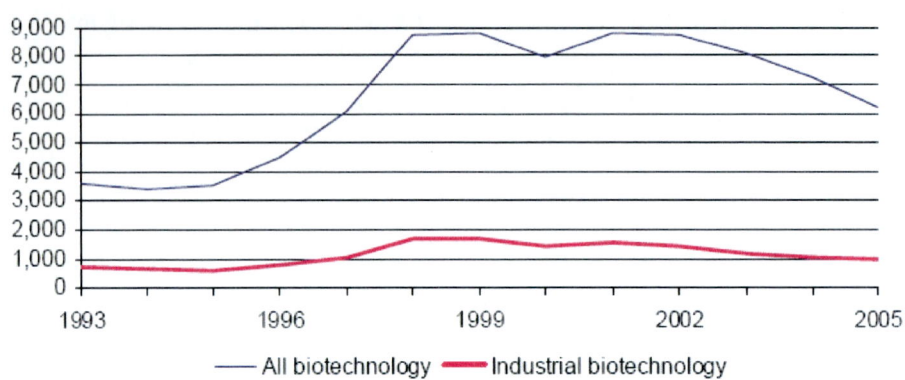

Source: USPTO, Utility Patents, 2007; USPTO, Patent Technology Center Groups 1630–1660, 2006.

Note: The first year that annual data for all biotechnology patents are available is 1993.

Figure 3.2. Biotechnology and industrial biotechnology patents granted yearly, 1993–2005.

The general decline in patent grants since the late 1990s, with some resurgence in 2006, can be attributed to various factors. The most important appears to be issues internal to the USPTO rather than declining levels of biotechnology innovation. Biotechnology patent applications increased by about 40 percent during the same period when patent grants were dropping.[10] Although the USPTO expanded its hiring of patent examiners during this period, high examiner attrition rates, as well as increasingly complex claims have translated into longer periods of patent pendency.[11] With more and more applications filed each year, the backlog of patent applications has continued to grow, from 308,056 applications awaiting examination in 2000 to 760,924 in 2007.[12]

Adelman and DeAngelis also note that more rigorous standards of review put into place in 2001 may have contributed to the decreasing number of grants.[13] Patent approval rates have been on the decline, from a 72 percent approval rate in 2000 to a 49 percent approval rate in 2006 apparently because of more rigorous review standards.[14]

Characteristics of Patent Ownership

Industrial biotechnology-related patents are predominantly owned by domestic and foreign corporations.[15] Such firms accounted for an average of 93 percent of all patents granted during the 1975–2006 period.[16] While the total number of patent granted to domestic corporations exceeds those granted to foreign corporations, the trends in grants to foreign and domestic corporations have been closely aligned (figure 3.3).

Spurred by Bayh-Dole, U.S. universities have played an increasingly prominent role in industrial biotechnology-related patenting (figure 3.4). U.S. universities obtained only 5 industrial biotechnology-related patents in 1975, a high of 235 patents in 1999, and 159 patents in 2006; an increase of more than 3,000 percent during the entire period. By contrast, patents issued to domestic corporations (excluding universities) grew by 716 percent during the period. Similarly, U.S. university patents, as a share of all patents, grew from 2.5 percent in 1975 to a peak of 15.1 percent in 1995, declining to 12.6 percent in 2006 (figure 3.4).

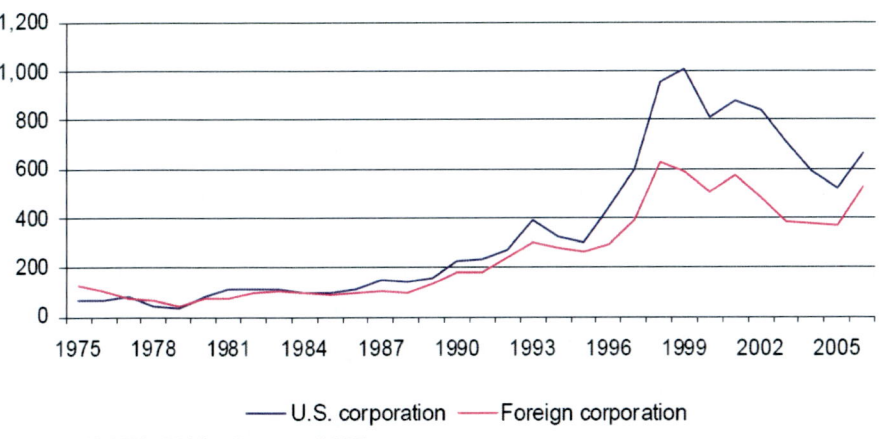

Source: USPTO, Utility Patents, 2007.

Figure 3.3. Industrial biotechnology patents granted yearly to U.S. and foreign corporations, 1975–2006.

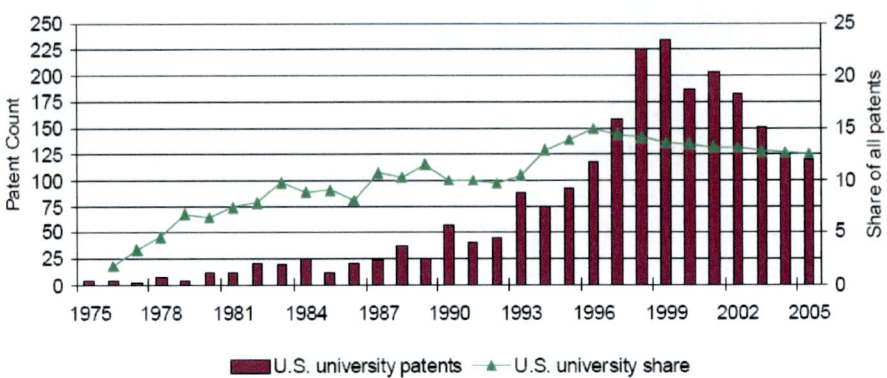

Source: USPTO, Utility Patents, 2007.

Figure 3.4. Industrial biotechnology patents granted yearly to U.S. universities and U.S. universites' share of all industrial biotechnology patents, 1975–2006.

By contrast, patents issued to the U.S. government grew more slowly, increasing from 4 patents in 1975 to a high of 25 in 1999, and reaching a high of 25 again in 2006, for an increase of 525 percent over the entire period (figure 3.5). However, the aggregate data have not been examined at the individual patent level to determine whether particular patents granted to universities or corporations were government funded. Thus, this increase in government-owned patents does not reflect the full extent of government involvement in industrial biotechnology discoveries.

A total of 2,978 owners (excluding individuals) obtained patents in industrial biotechnology-related classifications from 1975 through 2006.[17] The number of owners obtaining a patent in the relevant classifications each year grew steadily from 103 in 1975 to a peak of 593 in 1999, and declined to 534 in 2006.[18] The field also appears to be characterized by a steady stream of new participants. New owners, defined as an assignee that had not patented in any of the previous years in the period, grew from 103 in 1975, peaked at 200 in 1999, and declined to 151 new owners in 2006 (figure 3.6).[19]

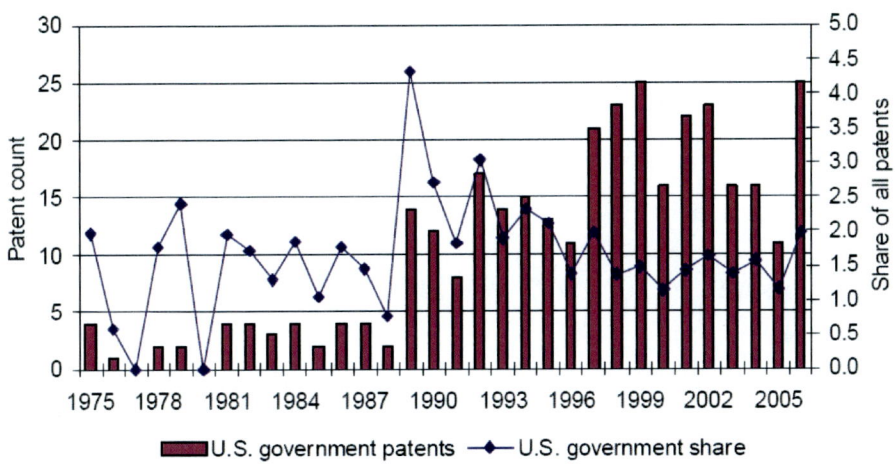

Source: USPTO, Utility Patents, 2007.

Figure 3.5. Industrial biotechnology patents granted yearly to the U.S. government and U.S. government share of all industrial biotechnology patents, 1975–2006.

This steady stream of new entrants is noteworthy because of the critical role that new entrants, and particularly start-up firms, have played in the development of the biotechnology industry.[20] A similar phenomenon may be occurring in industrial biotechnology.

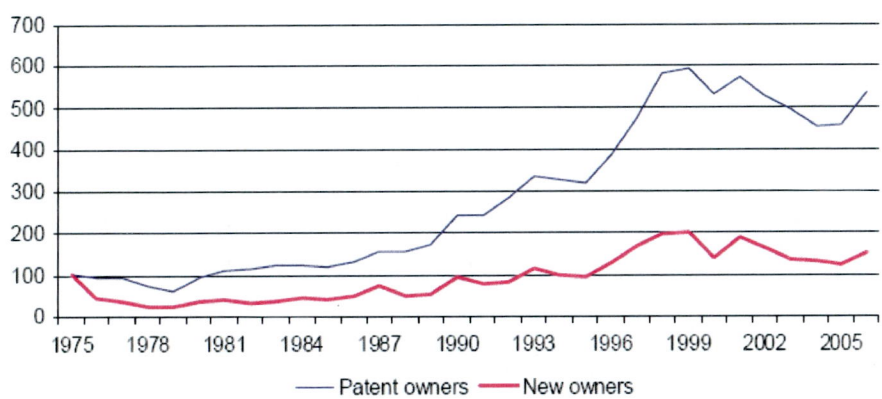

Source: USPTO, Utility Patents, 2007.
Note: Individual and unassigned patent owners excluded from calculations.

Figure 3.6. Industrial biotechnology patent owners and new owners, 1975–2006.

Table 3.1. Industrial biotechnology: Top patent owners, 2002–06

Name	Patents Granted 2002–06	Size	Country	Industry
Applera Corporation	146	Medium	United States	R&D
Novozymes A/S	103	Medium	Denmark	Enzymes
Millennium Pharmaceuticals, Inc.	99	Medium	United States	Pharmaceutical
Degussa Aktiengesellschaft	97	Large	Germany	Pharmaceutical
E. I. Du Pont De Nemours and Company	97	Large	United States	Chemical
Human Genome Sciences, Inc.	97	Small	United States	Pharmaceutical
Genencor International, Inc.	93	Medium	United States	Enzymes
Genentech, Inc.	86	Large	United States	Pharmaceutical
University of California	81	([a])	United States	University
Lexicon Genetics Incorporated	72	Small	United States	R&D

Source: USPTO Patent Full-Text Image Database; Bureau van Dijk, Orbis Companies Database.

Note: For purposes of this study, large companies are those with revenue above $3 billion; medium-sized companies are those with revenue in the range of $100 million to $3 billion; and small companies are those with revenue below $100 million.

[a] Not available.

Moreover, the number of patents held by each owner is relatively small; of the 2,978 owners, about one-half (1,451) hold only one patent. The number of patents held by the remaining owners ranges from 2 to 375, with a median value of 4. Looking at the most recent five-year period, table 3.1 identifies those owners that obtained the most patents, the number of patents obtained, and the owner's size and industry sector.[21]

The number of industrial biotechnology-related patents granted to the top 10 patenting owners in the last five years ranges from 72 to 146 (table 3.1). The number of patents issued during this entire five year period is far smaller than the number of patents issued on an annual basis to leading firms in the telecommunications and semiconductor fields. For example, in 2006 alone, the top patenting company in the United States, IBM, obtained 3,148 patents and number two, Samsung, obtained 2,725.[22] The leading owners obtaining industrial biotechnology-related patents cut across industry sectors including R&D, enzymes, and pharmaceuticals, and include the University of California. There were four medium-sized companies among the top 10. Several large and small firms were also represented, reflecting a diversity of fields and firm sizes.

COMMISSION QUESTIONNAIRE RESULTS

At the request of the Senate Committee on Finance, the Commission conducted a study of the competitive conditions affecting firms in the liquid fuel and chemical industries that are developing and adopting industrial biotechnology products and processes.[23] The study's methodology included a written questionnaire, part of which sought information about impediments to R&D and commercialization of industrial biotechnology products and processes. The questionnaire explicitly identified "patent barriers" as a potential impediment to R&D and commercialization efforts and asked firms to indicate the significance of this impediment on a scale ranging from not significant to very significant.

A total of 384 firms responded to the question about impediments to commercialization. Seventy-three percent of these firms rated patent barriers as a least significant impediment and 9 percent as very significant. Out of 28 factors identified by the Commission in the questionnaire as potential impediments to commercialization, "patent barriers" was one of the least likely factors to be rated by respondents as very significant; it was ranked 25th out of 28 factors in descending order of importance.[24]

Similarly, a total of 379 firms responded to the question about impediments to R&D. Seventy-one percent of these firms rated patent barriers as a least significant impediment and 10 percent as very significant. Out of 10 factors identified as potential impediments to R&D, the least likely factor to be identified as very significant was "difficulties accessing university technologies," "patent barriers" was ranked 9th out of the 10 factors, and "inability to establish alliances" was ranked 8th. Thus, intellectual property-related factors were not considered substantial impediments to R&D in industrial biotechnology.[25]

ENDNOTES

[1] The search methodology is set forth in app. A and search results are reported in USPTO, Utility Patents, 2007.

[2] The patents were in the following fields of industrial biotechnology: biofuels, biopolymers, enzyme technologies, chemical processes, and pharmaceuticals.

[3] USPTO, Overview of the U.S. Patent Classification System (USPC), June 2008.

[4] Because the USPC is a hierarchical system, included within the search were all classifications that were indented under a relevant subclass (i.e., all "children" of our subclasses). The classification numbering system is not sequential; there are gaps in numbers and indented classes may fall under subclasses with very different numbers.

[5] Chap. 4 describes particular emerging technologies in industrial biotechnology in more detail.

[6] USPTO, Patent Technology Center Groups 1630–1660, 2006; Adelman and DeAngelis, "Patent Metrics," 2007, 1687.

[7] USPTO, Patent Technology Center Groups 1630–1660, 2006. Adelman and DeAngelis identified a smaller number of biotechnology patents (52,039) granted during a shorter period, from January 1990 through December

2004, and excluded patents associated with agricultural biotechnology from their results. Adelman and DeAngelis, "Patent Metrics," 2007, 1741.
[8] U.S. government official, e-mail message to Commission staff, November 1, 2007.
[9] Adelman and DeAngelis, "Patent Metrics," 2007, 1691.
[10] Ibid., 1690.
[11] U.S. GAO, U.S. Patent and Trade Mark Office, September 2007.
[12] USPTO, Performance and Accountability Report Fiscal Year 2007, table 3, December 21, 2007.
[13] Adelman and DeAngeles, "Patent Metrics," 2007, 1689–90.
[14] Merritt, "Fixing the Patent Office," September 17, 2007.
[15] Patent ownership is based on identity of the first-named assignee of the patent on the date of the grant. While mergers and acquisitions, purchases and sales of intellectual property, and other corporate changes may impact ownership, these changes are not reflected in the aggregate data. See chapter 5 for suggestions for future research.
[16] The U.S. and foreign corporations ownership categories count predominantly corporate patents; however, patents assigned to other organizations such as nonprofit organizations and universities are also included in these categories. Although a data file was obtained from the USPTO separating U.S. universities from U.S. corporations, this separation was not available for foreign universities and foreign corporations. Accordingly, the authors left U.S. universities and corporations combined for purposes of comparing them to foreign corporations. The separate data for U.S. universities are described in the following discussion of university patenting trends.
[17] Individual and unassigned patents—i.e., those that could not be attributed to the foreign and domestic corporation, government, and university categories—were excluded from the count of patenting entities. The USPTO data file identifies a total of 740 individual and unassigned patents issued during the period. USPTO, Utility Patents, 2007.
[18] This calculation is based on the number of distinctly identified owners; for example, it does not take into account the parent-subsidiary relationship,

or other legal relationships, that may exist between patent owners, nor does it account for errors in the way names are recorded.

[19] All assignees were considered new entrants in 1975, year one of the calculation.

[20] See chap. 2.

[21] Information about firm size and industry sector was obtained from Bureau van Dijk, Orbis Companies Database, and company reports. This study defines large companies as those with annual revenues above $3 billion; medium are those with revenues in the range of $100 million to $3 billion; and small are those with revenues below $100 million.

[22] Wolters Kluwer Health, "IFI Patent Intelligence Announces 2007's Top U.S. Patent Assignees," January 14, 2008; Adelman and DeAngelis. "Patent Metrics," 2007, 1695.

[23] See USITC, Industrial Biotechnology, 2008.

[24] The most significant impediments included feedstock prices, lack of capital, and risk levels. USITC, Industrial Biotechnology, 2008, 3-2.

[25] The most significant impediments to R&D were lack of capital, U.S. regulatory requirements, and the limits of available technology. Ibid., 3–18.

Chapter 4

INDUSTRIAL BIOTECHNOLOGY AT THE TECHNOLOGY AND FIRM LEVEL

This chapter looks at two emerging sectors of industrial biotechnology at the technology and firm level: cellulase-related enzymes, which are used to break down cellulosic biomass as part of the production of biofuels and bio-based chemicals, and enzyme technologies that enable the production of bio-based plastics such as polyhydroxyalkenoates (PHA) and polylactides (PLA).

At the technology level, USPC classifications that appeared to match well with these sectors were selected and patents issued in these classifications in the last 10 years were identified through USPTO's online database.[1] Foreign and domestic corporations are the dominant patent owners in all classifications, most owners hold a relatively small number of patents, and new owners are steadily entering these emerging fields. At the firm level, top companies patenting in these classifications were selected to profile. Profiles are provided for two companies developing and/or using cellulase-related enzymes, Novozymes and Verenium, and two companies focused on the production of bio-based plastics, Metabolix and Cargill (through its NatureWorks LLC subsidiary).

NEW ENZYME TECHNOLOGIES

The use of enzymes in an industrial process is a marker of industrial biotechnology.[2] The development of new enzymes, including through the production and purification of enzymes from genetically modified organisms, is a major driving force in the commercialization of industrial biotechnology products and processes. In addition to the production of novel products, enzymes can be used to make fuels and chemical intermediates in more sustainable, environmentally friendly ways. For example, cellulosic ethanol is believed to have environmental and food security advantages over corn-based ethanol because it can be made from a variety of nonfood feedstocks and involves the use of the entire crop rather than just the kernel.[3]

Cellulases, enzymes that break down the cellulose and hemicellulose that make up cellulosic biomass and convert them into sugars, are considered critical to the cost-effective production of cellulosic ethanol.[4] Many cellulase-related patents are classified under USPC class 435: "Chemistry: molecular biology and microbiology" and subclass 209: "Acting on a beta-1,4-glucosidic bond (e.g., cellulase)."[5] Cellulase-related patents classified under 435/209, and the activities of new enzyme technology companies, are profiled below.

Cellulase-related Patents

The USPTO issued 277 patents with 435/209 as their primary or secondary classification during the 1997–2007 period. Foreign corporations are the dominant patent owners in this classification, with 63 percent of all issued patents compared to 15 percent for domestic corporations, followed by U.S. and foreign educational institutions and governments (figure 4.1).

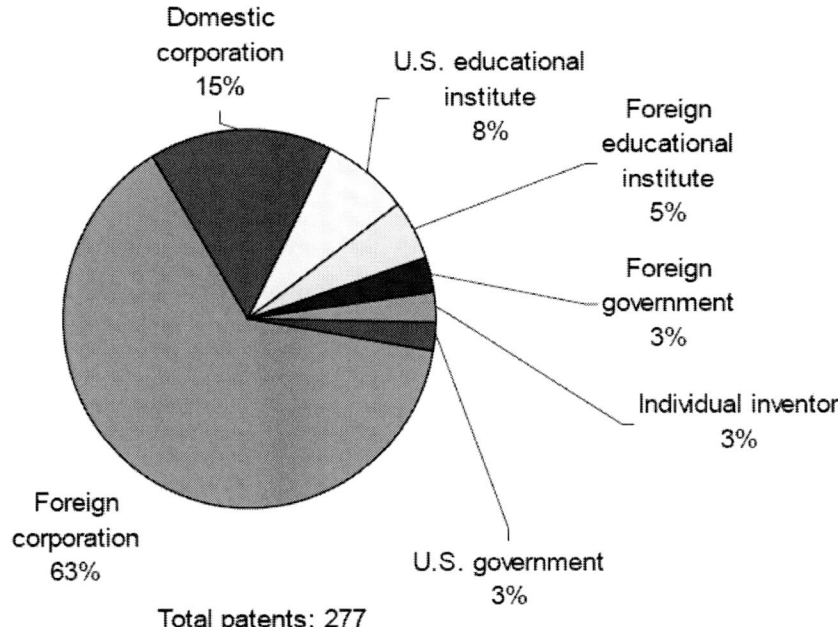

Source: USPTO, USPTO Patent Full-Text and Image Database.

Figure 4.1. Patents granted by owner type in classification 435/209, 1997–2007.

Danish owners hold the largest share of patents in this classification, 41 percent of issued patents. U.S. owners hold 27 percent, followed by Japan (6 percent), Canada (6 percent), Germany (4 percent), and the Netherlands (4 percent). The Danish-owned patents are held predominantly by two companies, Novo Nordisk A/S and Danisco A/S.[6] Most of the other owners hold a small number of patents; of the 109 different entities granted patents in this classification, 80 of them obtained only one patent during the 1997–2007 period. For the other 29 entities, the number of patents received ranges between 2 and 68, with a median value of 3.[7]

The top patent owners in the class, and their country of origin and size, are identified in table 4.1.[8] While this class is dominated by large companies Novo Nordisk A/S and Danisco A/S, smaller companies also own multiple patents. One such small company, Verenium, is the third largest patent holder in the class (see profile in the next section). Two of the top companies are education-related, the University of Georgia Research Foundation and the University of British Columbia.

Moreover, the number of "new" owners—that is, owners receiving their first patent in this time period—is more than one-half of the total number of owners. New owners have been steadily entering the field (figure 4.2). The HerfindahlHirschman Index (HHI), a measure of market concentration, can be used to assess whether patenting in this technology area is controlled by a small number of firms or is open to new entrants. [9] Here, an HHI of 0.09 suggests that control of the cellulase-related patent field is unconcentrated, despite two companies having a large presence. The existence of many firms patenting in this class, with most having only one patent, lowers the index and keeps the market from being deemed highly, or even moderately controlled.

Table 4.1. Top patent owners in classification 435/209, 1997–2007

Company Name	# of	Company	Country
Novo Nordisk A/S	68	Large	Denmark
Danisco A/S	42	Large	Denmark
Verenium Corporation	7	Small	United
University of Georgia	7	Small	United
Lockheed Martin Corporation	6	Large	United
Iogen Corporation	6	Small	Canada
Novartis AG	5	Large	Switzerland
Koninklijke DSM N.V.	4	Large	Netherlands
Meiji Seika Kaisha Ltd.	4	Medium	Japan
University of British Columbia	4	Medium	Canada

Source: USPTO, USPTO Patent Full-Text and Image Database; Bureau van Dijk, Orbis Companies Database.

Note: For purposes of this study, large companies are those with revenue above $3 billion; medium-sized companies are those with revenue in the range of $100 million to $3 billion; and small companies are those with revenue below $100 million.

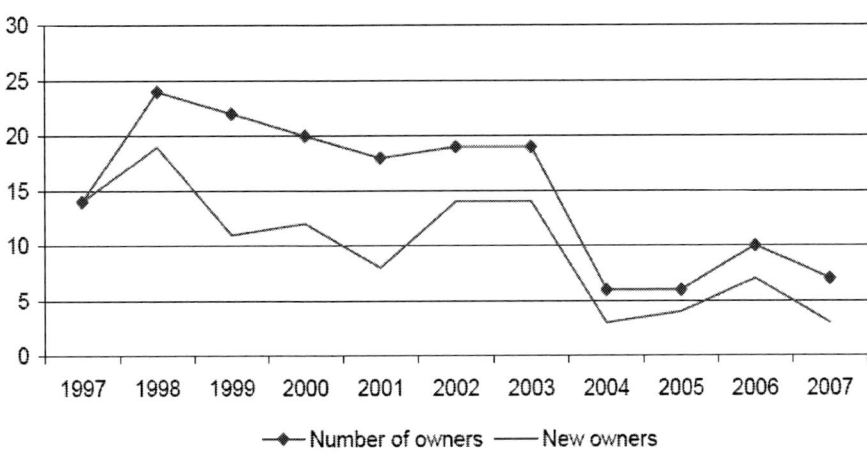

Source: USPTO, USPTO Patent Full-Text and Image Database.
Note: "New owners" counts the number of assignees in that year that are receiving their first patent in 435/209 in the 1997–2007 period.

Figure 4.2. Patent owners and new owners in classification 435/209, 1997–2007.

The use of the HHI as an indicator of control over patent rights relies upon a number of assumptions that merit caution in the interpretation of the results. First, it assumes that the control or ownership of patents in the group does not change significantly after the patent is issued. The Commission's recent report on industrial biotechnology, however, indicates that purchase, sale, and licensing transactions are increasing.[10] This use of the HHI also assumes that each of the patents in the group is of equal value; however, empirical research suggests that large firm patents are more valuable than those of small firms.[11] Moreover, some of the patents in the group likely are not relevant because they have been abandoned or invalidated by legal challenge or because of misclassification. These caveats suggest areas in which further research is warranted to reduce the "noise" associated with patent data.[12]

Enzyme Technologies Profile: Novozymes

Novozymes, a leader in the development of new enzyme technologies, relies on a strong patent portfolio to protect R&D investments, secure the freedom to operate in a particular patented space, and protect and increase market share.[13] Novozymes continues to develop new and improved products through growth in the company's R&D expenditures, participation in U.S. government research projects, and collaborations with other firms.

Company Overview

Novozymes is a producer of enzymes and micro-organisms for pharmaceutical and industrial uses.[14] The company was founded in Denmark as a medical firm in 1925 and began producing enzymes for industrial use (for the softening of leather) in 1941.[15] Novozymes has production and research facilities in Australia, Brazil, China, Denmark, Sweden, the United States, and the United Kingdom, and research facilities in India.[16] While the largest share of Novozymes' revenue currently comes from detergent enzymes, in the last few years, sales of technical enzymes, such as those that convert starch to sugars for the production of ethanol, have grown at a faster pace and are now almost equal to the company's sales of detergent enzymes.[17]

In 2007, Novozymes had revenue of $1.54 billion, net income of $213 million, and about 4,700 employees (table 4.2).[18] Novozymes' strong revenue and net income growth over the period have enabled growing R&D expenditures, which increased from $138.0 million in 2001 to $202.6 million in 2007. The strategy for the future growth of the company is to expand the market for enzymes by producing new products, and new applications for current products, through substantial R&D efforts.[19]

Novozymes also is involved in strategic alliances with government and industry that focus on industrial biotechnology and cellulosic ethanol. For example, in 2001, Novozymes received a multi-year U.S. Department of Energy (DOE) contract for $17.8 million in matching funds with the goal of achieving a substantial cost reduction in cellulase enzymes. More recently, in 2008, Novozymes obtained an additional DOE award to improve the performance of its most advanced enzyme system for the production of cellulosic ethanol.[20] Novozymes has also partnered with China Resources Alcohol Corp. to produce biofuels from cellulose for the Chinese market.[21]

Table 4.2. Novozymes: Financial data, 2001–07

	2001	2002	2003	2004	2005	2006	2007
Revenues (million $)	1,081	1,160	1,190	1,225	1,286	1,398	1,539
R&D expenditures (million $)	138.0	145.2	152.5	158.8	161.4	179.2	202.6
Government grants (million $)	7.7	9.0	8.3	4.1	1.8	1.8	1.6
Net income (million $)	112.6	131.1	147.8	155.5	174.7	185.1	213.3
Employees (number)	3,349	3,629	3,814	3,928	4,023	4,272	4,684

Source: Bureau van Dijk, Orbis Companies Database.
Note: Detailed financial information available from Orbis beginning in 2001.

Novozymes' Patent Portfolio

Novozymes has obtained an estimated 334 patents, and has pending an estimated 447 patent applications with the USPTO for the 1997–2007 period (figure 4.3).[22] Novozymes' patent filings increased sharply from 1997 to a peak in 2001 and declined thereafter.

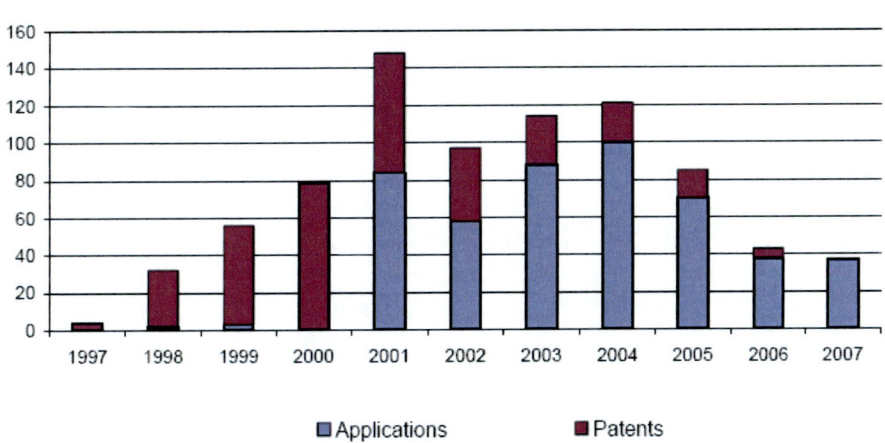

Source: USPTO, USPTO Patent Full-Text and Imgage Database.
Note: The USPTO began publishing patent applications in 2001.

Figure 4.3. Novozymes: U.S. patents and applications, by application date, 1997–2007.

The most common class for patents granted to Novozymes in these years was 435: "Chemistry: molecular biology and microbiology"; nearly 69 percent of all U.S. patents granted to Novozymes were in this class. The next most common classes were 426: "Food or edible material: processes, compositions, and products" (10 percent), and 424: "Drug, bio-affecting and body treating compositions" (6 percent). The most common subclass under class 435 for Novozymes' patents was 435/69.1: "Recombinant DNA techniques included in method of making a protein or polypeptide." This classification is important to industrial biotechnology because the activities include genetic engineering to produce enzymes with new functionality or to produce existing enzymes more efficiently.[23]

Novozymes and Genencor (and their parent companies) are leading patent owners in enzyme technologies, and have been involved in intellectual property disputes relating to industrial biotechnology. In March 2005, Novozymes sued Genencor for patent infringement over alpha amylase enzymes used in the production of ethanol. Novozymes prevailed and obtained a settlement that included a $15 million cash payment by Genencor.[24] Not surprisingly, Novozymes considers patent rights to be extremely important to the company because they protect investment in R&D, secure the freedom to operate in patented technology areas, reduce the operational freedom of competitors, and help maintain and sometimes increase market share.[25]

Cellulosic Ethanol Profile: Verenium

The Verenium profile highlights the importance of patents to the transfer of technology from the university to the small firm setting for further development, and then to larger firms with the resources needed to move the technology toward commercialization. Patents enable the strategic alliances and government grants necessary to continue R&D and innovative activities during the pre- commercialization phases when revenues from product sales are limited.

Company Overview

Verenium was formed in 2007 through the merger of Diversa, a company developing and producing specialty enzymes, and Celunol, a company focused on cellulosic ethanol technology.[26] A key asset of the merged business is patented technology which was first discovered by Dr. Lonnie Ingram and colleagues at the University of Florida for breaking down cellulosic biomass to produce cellulosic ethanol.[27] Verenium has two separate business units: the biofuels business unit, which is developing integrated cellulosic ethanol production capabilities at a pilot and demonstration plant in Jennings, Louisiana, and the specialty enzymes unit, which focuses on alternative fuels, industrial processes, and animal nutrition and health.[28]

Verenium (and its predecessor companies) have continuously been in a negative cash flow position. R&D expenditures increased by nearly 20 percent from 2001 through 2007 while revenues declined (table 4.3). Verenium was in a particularly precarious financial position at the end of 2007, with mounting losses and "substantial doubts" about its ability to continue.[29] In August 2008, however, the company's fortunes improved with the announcement of a new strategic alliance with BP.

Table 4.3. Verenium: Financial data, 2001–07

	2001	2002	2003	2004	2005	2006	2007
Revenues (thousand $)	9,858	3,927	2,286	1,767	2,011	2,307	3,877
R&D expenditures (thousand $)	48,228	50,096	70,695	73,405	72,751	50,033	57,727
Government grants (thousand $)	910	1,047	3,923	10,241	10,079	3,317	2,717
Net loss (thousand $)	-15,664	-29,633	-57,696	-33,425	-89,718	-39,271	-107,585
Employees (number)	276	279	360	322	287	287	280

Source: Bureau van Dijk, Orbis Companies Database.
Note: Detailed financial information available from Orbis beginning in 2001.

The BP alliance focuses on accelerating the development and commercialization of cellulosic ethanol technologies. During the first phase, BP will provide $90 million over 18 months for participation rights in a 50/50 partnership and the co-funding of technology development. Intellectual property plays a central role in the alliance. Verenium and BP have formed a special purpose entity into which each company's current technologies will be licensed, and which will own the new technologies that the companies plan to jointly develop to move cellulosic ethanol into commercialization.[30]

Government grants have also fueled Verenium's operations. Since 2005, the company has received government contracts and grants totaling $45 million from the U.S. Departments of Defense and Energy, and the National Institutes of Health. In March 2007, for example, Verenium was awarded up to $5.3 million by the DOE to continue enhancement of the organisms used in the cellulose-toethanol process, including improvements to the fermentation technologies licensed to Verenium from the University of Florida.[31]

Verenium's Patent Portfolio

Verenium reports that it owns 347 patents worldwide, has over 400 patent applications pending, and has licensed technologies from over 100 patents and applications. Verenium also holds exclusive worldwide licenses to use, develop, and commercially exploit the ethanol production patent estate of the University of Florida Research Foundation, including 15 U.S. and 54 foreign patents, 7 pending U.S. and 52 pending foreign patent applications, and other related intellectual property.[32] The University of Florida technologies include the seminal discovery of genetically modified E. coli bacteria that can convert the pentoses and hexoses (C5 and C6 sugars that result from the breakdown of hemicelluloses) into ethanol.[33]

The USPTO online database identifies 104 patents and 87 pending applications filed by Verenium, or its predecessors Diversa and Celunol, during the 1997–2007 period. Patent activity grew steadily to a peak in 2002, and has been on the decline since then (figure 4.4.).

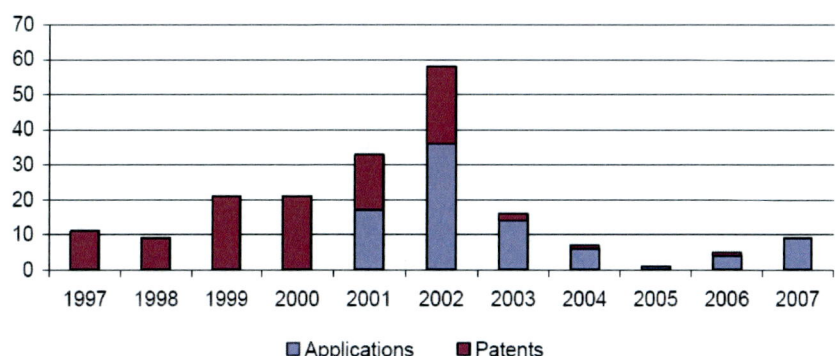

Source: USPTO, USPTO Patent Full-Text and Image Database.
Note: The USPTO began publishing patent applications in 2001.

Figure 4.4. Verenium: U.S. patents and application, by application date, 1997–2007.

Most of the patents granted to Verenium were in class 435: "Chemistry: molecular biology and microbiology," and nearly 50 percent of these were in subclasses 4, 6, and 69.1, with about one-half classified in subclass 6. Subclass 4 covers measuring or testing processes involving enzymes or micro-organisms and 6, which falls under this subclass, covers testing processes that include nucleic acid. Verenium reports that its proprietary technologies include capabilities for sample collections from microbial populations, generation of DNA libraries, screening of the libraries using high-throughput methods, and optimization based on gene evolution technologies; these screening technologies fall within the subclasses 4 and 6.[34]

Bio-based Plastics

Bio-based plastics are another emerging sector of industrial biotechnology in which patents play a central role. Polymers made from renewable resources and produced using enzymes and micro-organisms are increasing in prominence as technological advances and rising crude petroleum prices make them more technically and economically competitive with petroleum-based plastics. Many bio-based plastics also offer environmental benefits, such as the ability to biodegrade, that increase their appeal to consumers.[35] Some of the most innovative industrial biotechnology R&D is focused on bioplastics synthesized through fermentation of polysaccharides to create PHA and PLA.[36]

PHA is produced via bacterial fermentation of sugars or lipids. PHA can be used in a wide variety of applications including coated paper, film or bags, thermoformed and molded goods, and as building blocks for applications such as solvents and chemical intermediates.[37] Many patents related to PHA are classified under class 435: "Chemistry: molecular biology and microbiology," and subclass 135: "Carboxylic acid ester: processes wherein the product synthesized contains an ester group."[38]

PHA-Related Patents

The USPTO issued 308 patents with 435/135 as their primary or secondary classification during the 1997–2007 period. As with cellulase-related patents, foreign corporations own most of the patents in this class, followed by domestic corporations and U.S. educational institutions (figure 4.5). The single country with the largest share of patents in this classification is the United States; 41 percent of all patents in this classification were issued to U.S. owners (corporations, universities, and government). Japan holds the second largest country share with 25 percent, followed by Germany (9 percent), Switzerland (4 percent), and India (3 percent).

The top patent owners in the class and the size and location of the company are identified in table 4.4. The patents held by second-ranked Metabolix, a small U.S. firm, likely exceed those of the first-ranked company, Canon. In addition to the 18 patents identified in table 4.4, Metabolix acquired the PHA-related patents of Monsanto, the next largest patent holder in the class, in 2001.[39] While more than one-half of the largest patent owners in this classification are large companies (6 of 11), both small and medium-sized entities are well represented. The companies patenting in this class hold a relatively small number of patents. Of the 253 patents held by foreign and domestic corporations, 72 entities held 1 patent. The median for the remaining companies was 3 patents, with a range between 2 and 22 patents.

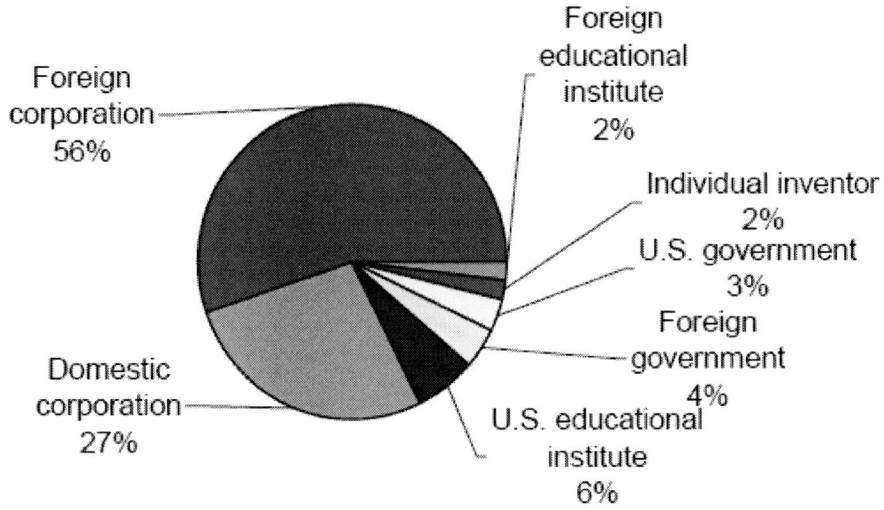

Total patents: 308

Source: USPTO, USPTO Patent Full-Text and Image Database.

Figure 4.5. Patents granted by owner type in classification 435/135, 1997–2007.

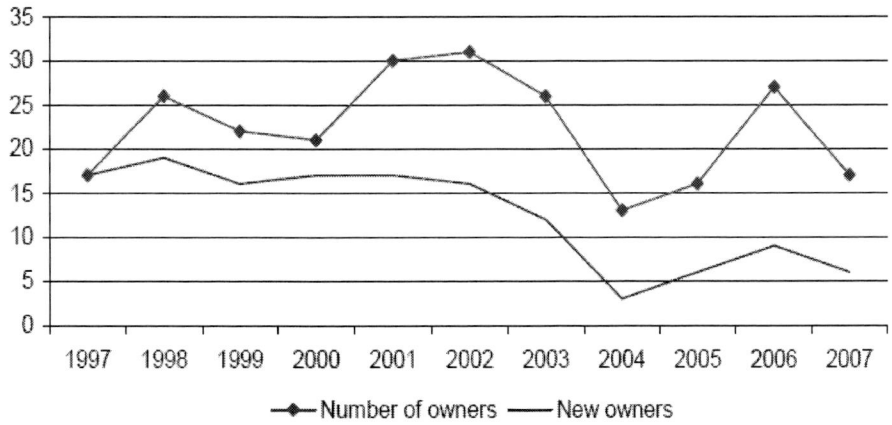

Source: USPTO, USPTO Patent Full-Text and Image Database.
Note: "New owners" counts the number of assignees in that year who are receiving their first patent in 435/135 in the 1997–2007 period.

Figure 4.6. Patent owners and new owners in classification 435/135, 1997–2007.

Table 4.4. Top patent owners in classification 435/1 35, 1997–2007

Assignee Name	# of patents Company size	Country
Canon Kabushiki Kaisha	22 Large	Japan
Metabolix Inc.	18 Small	United States
Monsanto Company	10 Large	United States
Council of Scientific and Industrial Research	9 (a)	India
U.S. government	8 (a)	United States
E. I. du Pont de Nemours and Company	7 Large	United States
Suntory Limited	6 Large	Japan
The Procter & Gamble Company	6 Large	United States
Novo Nordisk A/S	6 Large	Denmark
Daicel Chemical Industries, Ltd.	6 Medium	Japan
HENKEL FAMILY	6 Medium	United States

Source: USPTO, USPTO Patent Full-Text and Image Database; Bureau van Dijk, Orbis Companies Database.

Note: For purposes of this study, large companies are those with revenue above $3 billion; medium-sized companies are those with revenue in the range of $100 million to $3 billion; and small companies are those with revenue below $100 million.

[a] Not available.

From 1997 through 2007, the number of entities (corporations, governments, and universities) obtaining patents peaked at 31 in 2002. The number of new entities granted a patent for the first time during this period followed a sporadic trend, peaking with 19 new owners entering the field in 1998 and declining in recent years (figure 4.6). The HHI of 0.02 indicates that control of the patent rights in the PHA-related technology is unconcentrated.[40]

PHA Profile: Metabolix

Metabolix has its PHA roots in technology patented by scientists at the Massachusetts Institute of Technology (MIT). The company has used its strong patent portfolio to attract strategic alliance partners from government and industry that bring the capital necessary to move from R&D to commercialization of PHA bio-based plastics.

Company Overview

Metabolix is a small company focused on the development and commercialization of environmentally sustainable alternatives to petrochemical- based plastics, fuels, and chemicals.[41] The company was formed in 1992 by scientists from MIT who identified the key genes required for the biosynthesis of PHA and invented and patented the first transgenic system for its production. To fully capture the PHA opportunity, Metabolix later acquired a patent estate related to the production of PHA from a wild-type bacterial strain that had been owned by Monsanto. Metabolix also has an active R&D program focused on the development of a biomass refinery system using switchgrass to co-produce both PHA and biomass to be used for fuel power production.[42]

Metabolix is currently developing a proprietary, large-scale, microbial fermentation system for the production of PHA (which Metabolix brands as Natural Plastic). The company has formed a joint venture with Archer Daniels Midland (ADM), one of the largest U.S. producers of biofuels and agricultural products, under the name of Telles that will operate the first commercial plant to produce PHA. The new plant will be located in Clinton, Iowa, next to an ADM ethanol production facility, and is scheduled to come online in 2009. The plant is expected to produce up to 110 million pounds of bioplastics annually.[43] Telles will reportedly employ a "systems approach" to bioplastic production, integrating sophisticated biotechnology with advanced industrial practice from the initial genetic engineering to end product.[44]

Table 4.5. Metabolix: Financial data, 2001–07

	2001	2002	2003	2004	2005	2006	2007
Revenues (thousand $)	(a)	(a)	6	14	-591	1,467	5,941
R&D expenditures (thousand $)	6,309	4,409	6,204	5,427	5,980	11,235	19,901
Government grants (thousand $)	(a)	(a)	(a)	3,189	2,433	1,828	879
Net loss (thousand $)	-8,936	-5,188	-6,641	-5,055	-7,625	-16,062	-27,875
Employees (number)	(a)	(a)	(a)	(a)	43	59	77

Source: Bureau van Dijk, Orbis Companies Database.
Note: Detailed financial information available from Orbis beginning in 2001. [a]Not available.

Metabolix has made substantial R&D investments in its PHA products and its biomass refinery system. One of the primary goals of the R&D program is to "develop and acquire competitive intellectual property and know-how in biobased plastics, fuels, and chemicals that define [the company] as the leader in the field."[45] To carry out its R&D program, expenditures have increased by 221 percent from $6.2 million in 2003 to $19.9 million in 2007 (table 4.5). Revenues, while growing from essentially zero to $5.9 million in 2007, are far from covering R&D expenses.

Metabolix receives funding from the U.S. government and its alliance partners, enabling it to maintain and even expand its research and commercialization efforts, notwithstanding its negative cash flow position. For example, in 2001 the DOE awarded Metabolix $7.5 million to develop a co-production process for Natural Plastic with energy crops and in 2003, the U.S. Department of Agriculture awarded Metabolix an additional $2 million for this program. Then in 2007, the U.S. Department of Commerce awarded Metabolix $2 million for the ongoing development of a commercially viable platform for producing biobased chemicals from agricultural products.[46]

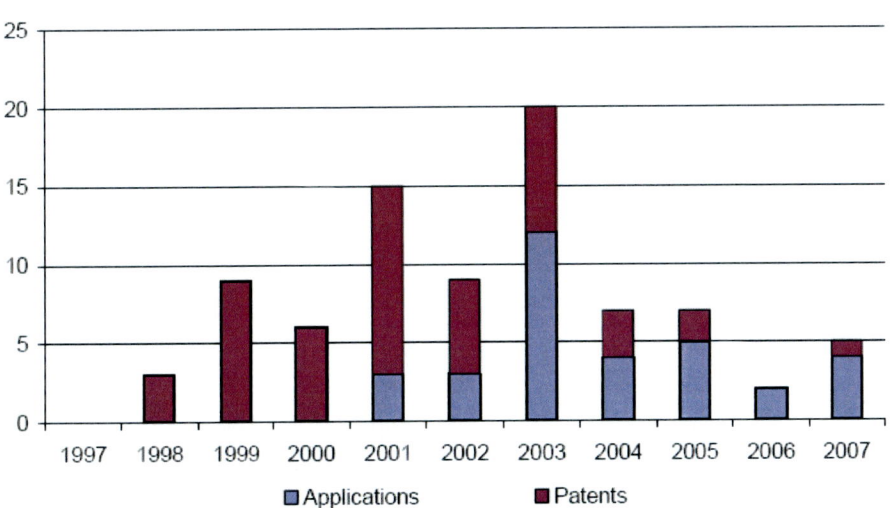

Source: USPTO, USPTO Patent Full-Text and Image Database. Note: The USPTO began publishing patent applications in 2001.

Figure 4.7. Metabolix: U.S. patents and applications, by application date, 1997–2007.

Metabolix's Patent Portfolio

Metabolix reports that its "continued success depends in large part on its proprietary technologies," and that it relies on patent, copyright, trademark and trade secret laws, and confidentiality agreements to establish and protect its intellectual property rights.[47] Metabolix reportedly owns over 340 issued patents and 120 patent applications worldwide, and holds licenses to use technologies in another 60 issued patents and 30 patent applications worldwide. Licensed patents and applications include 13 U.S. patents and their foreign counterparts issued to MIT scientists and exclusively licensed to Metabolix.[48]

The USPTO online database identifies 50 patents and 32 pending applications filed by Metabolix during the 1997–2007 period. Patent activity peaked in 2003 and has been trending downward since then, with a small rebound in 2007 (figure 4.7).

Metabolix patents cover polymers, genes, vectors, expression systems in plants and microbes, devices, coatings, films, as well as methods of manufacture and use.[49] They are grouped around a small number of technology areas in class 435: "Chemistry, molecular biology and microbiology," with the most active subclass being the PHA-related classification 435/135. The combination of patents owned and exclusively licensed by Metabolix gives it a powerful position in the PHA market. The patents protect methods of PHA isolation, purification, and processing, preferred metabolic pathways for copolymer production, and several novel compositions and applications. Most fundamentally, according to researchers for the Environmental Protection Agency, they give Metabolix "exclusive rights to the genes within the PHA biosynthesis pathway, as well as the use of the genes in any combination for the preparation of PHA."[50]

Given the strong position that the Metabolix patents provide, it is perhaps not surprising that one of the company's competitors, Procter & Gamble—which has developed its own PHA polymers and licensed them to Kaneka Corporation—is challenging the validity of one of the foundational MIT patents licensed to Metabolix. The action is pending in the federal patent court in Munich, Germany. Metabolix maintains that the action is without merit but notes that the ability to obtain and successfully enforce its intellectual property in the United States and abroad is critical to the company's future success.[51]

PLA-Related Patents

Another bio-based plastic, PLA, is a compostable polyester derived from lactic acid, which is made from agricultural byproducts through the processes of fermentation, distillation, and polymerization of plant sugars. A number of finished products can be made from PLA, including structural supports and drug delivery systems in medical devices, tough fibers, and molded or extruded consumer products such as food packaging.[52] Many PLA-related patents are classified under class 435: "Chemistry: molecular biology and microbiology," and subclass 139: "Lactic acid: processes wherein the product synthesized is an acid or salt form of alpha-hydroxy propanoic acid."[53]

The USPTO issued 69 patents with 435/139 as their primary or secondary classification during the January 1997–December 2007 period. Foreign and domestic corporations hold similar numbers of patents and together account for 85 percent of the total patents classified under 435/139 (figure 4.8). The United States holds the largest country share of patents with 51 percent, followed by Japan with 11 percent, Denmark with 8 percent, and France, Germany, and the Netherlands all with 6 percent.

The top patent owners in the class, and the size and location of the patenting companies are identified in table 4.6. A large, privately held U.S. company, Cargill, is the top patent holder in the class. Three of the top patent holders are located in the United States with the remaining companies in Japan and Europe. Most of the top patent owners (8 of 12) are large businesses. Most entities with patents in this class hold only one patent. The median for the remaining entities is two patents, with a range between two and seven patents. The HHI, estimated at 0.04, suggests that control of the patent rights in this technology area is unconcentrated. Overall, the class is characterized by a small number of patents spread over a small number of participants.[54]

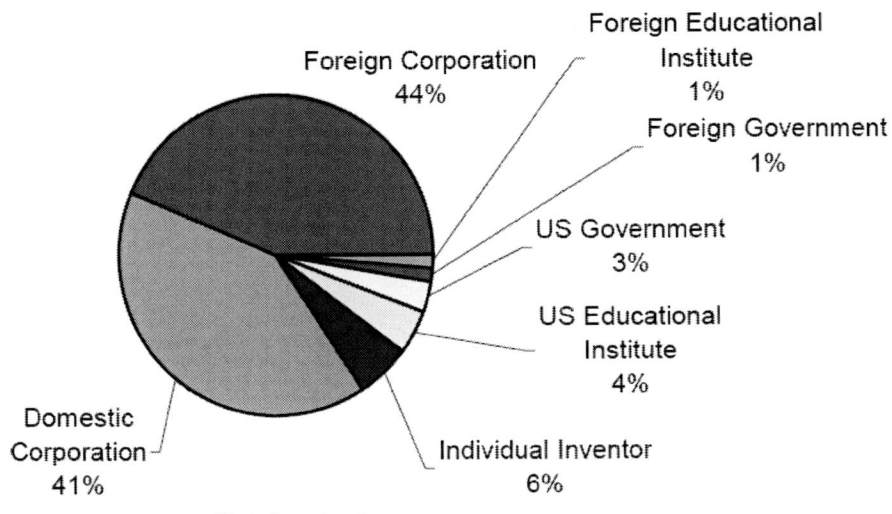

Source: USPTO, USPTO Patent Full-Text and Image Database.

Figure 4.8. Patents granted by owner type in classification 435/139, 1997–2007.

Table 4.6. Top patent owners in classification 435/139, 1997–2007

Assignee Name	# of patents	Company size	Country
Cargill & MacMillan Families	7	Large	United States
A. E. Staley Manufacturing Co.	4	Small	United States
Archer Daniels Midland Co.	4	Large	United States
CSM NV	3	Large	Netherlands
Bayer AG	2	Large	Germany
Canon Inc	2	Large	Japan
Chr. Hansen Holding A/S	2	Medium	Denmark
Danisco A/S	2	Large	Denmark
Mitsubishi Rayon Co., Ltd.	2	Large	Japan
Roquette Freres	2	([a])	France
Tate & Lyle Public Limit Company	2	Large	United Kingdom

Source: USPTO, USPTO Patent Full-Text and Image Database; Bureau van Dijk, Orbis Companies Database.

Note: For purposes of this study, large companies are those with revenue above $3 billion; medium- sized companies are those with revenue in the range of $100 million to $3 billion; and small companies are those with revenue below $100 million.

[a] Roquette Freres is a family-owned, privately held company. No financial information is publicly available.

PLA Profile: Cargill

The PLA-related patent portfolio of Cargill and its subsidiary, NatureWorks, has attracted strategic alliances and government funding to advance the development and commercialization of bio-based plastics. NatureWorks has a strong competitive position with the only world-scale PLA plant in the United States, due, at least in part, to the strength of patent portfolios.[55]

Company Overview

Cargill is a large, privately held business with roots dating back to the 1800s in grain storage and trading.[56] Today, Cargill has about 158,000 employees in 66 countries and provides food, agricultural, and risk management products and services. Cargill reported total revenue of $88.2 billion for 2007, an increase of about 17 percent over 2006. The company operates under five business segments: agricultural services; food ingredients and applications; industrial; origination and processing; and risk management and financial. The company's subsidiaries include NatureWorks, LLC.[57] NatureWorks arose out of Cargill's 1989 patented discovery of a way to more efficiently make PLA from corn.[58] Cargill formed a joint venture with Dow in 1997 to further develop PLA, and in 2002 Cargill opened a PLA production facility in Blair, Nebraska. Dow exited the venture in 2004, and until 2007, the facility was operated solely by NatureWorks as a wholly owned Cargill subsidiary.[59] In 2007, Cargill and the Japanese company Teijin formed a 50/50 joint venture to operate NatureWorks. With more than 200 employees, NatureWorks has the capacity to produce 300 million pounds of polymer annually.[60]

Cargill has made substantial R&D investments in PLA, investing as much as $200 million during the 1990s alone.[61] These efforts have been supported by government funding in several programs. For example, in 1994, Cargill received an Advanced Technology Program award from the National Institute of Standards and Technology to develop the methodology for improving the performance characteristics of PLA to make it more competitive with petroleum-based plastics.[62] In 2000, Cargill Dow received additional funding from DOE to continue its R&D related to PLA.[63] According to Cargill, government funding accelerated commercialization and enabled the PLA project to "stay alive" at critical times.[64]

Cargil's Patent Portfolio

Cargill holds an estimated 217 patents and 175 pending applications at the USPTO for the 1997–2007 period. Cargill's patent filings increased steadily until 2002 and have been declining since then (figure 4.9). Cargill's USPTO patent holdings are spread across 44 discrete classes; these holdings reflect the diversity of products and services in which the company is active. Cargill's portfolio differs from those of the other profiled companies, which are more focused on enzyme technologies. Patents related to industrial biotechnology account for a relatively small portion of Cargill's total patent portfolio.[65]

Cargill's R&D has, however, yielded a valuable patent portfolio protecting its PLA-based polymer synthesis processes in the United States and other major markets.[66] Continuous production processes and technologies have been important to achieving cost-effective production. Recent Cargill patent applications covering the fermentation of lactic acid focus on making PLA production more efficient and sustainable by identifying processes in which the culture is capable of growing at lower pH levels and with enhanced recovery of lactic acid.[67] NatureWorks also holds a suite of patents and applications covering PLA processes and product discoveries.[68] With the only world-scale PLA plant in the United States, NatureWorks has a strong competitive position due, at least in part, to the strength of the Cargill and NatureWorks patent portfolios.[69]

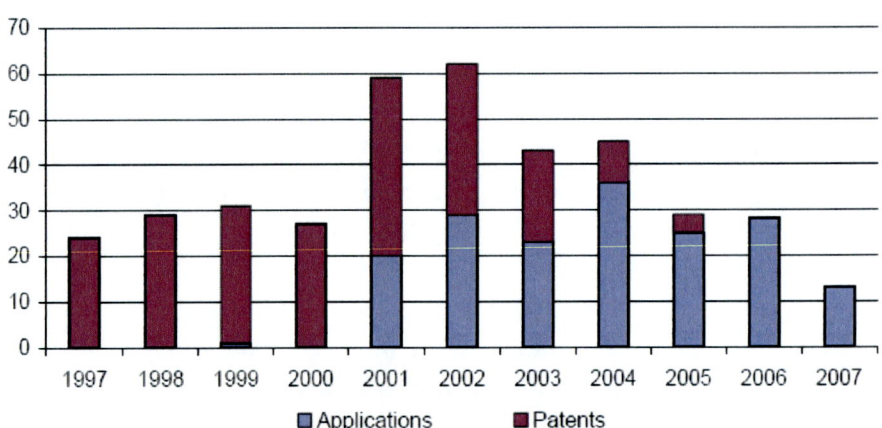

Source: USPTO, USPTO Patent Full-Text and Image Database.

Figure 4.9. Cargill: U.S. patents and applications, by application date, 1997–2007.

ENDNOTES

[1] The USPTO's online database was used instead of the aggregate data provided in the Utility Patents report to profile these emerging fields because the online database provides access to application data.

[2] USITC, Industrial Biotechnology, 2008, 1–3.

[3] Verenium, "Next Generation Cellulosic Ethanol," undated (accessed October 1, 2008).

[4] Knauf and Moniruzzaman, "Lignocellulosic Biomass Processing," 2004, 148. Cellulases also have applications in textiles and food production, as well as other industries.

[5] In the USPC classification hierarchy, this subclass falls below 183: "enzyme (e.g., liases, etc.), proenzyme; compositions thereof; process for preparing, activating, inhibiting, separating, or purifying enzymes," and indented subclasses 195: "hydrolases" and 200: "acting on glycosyl compound." USPTO Web Site "Main Classification Menu," undated (accessed August 11, 2008).

[6] Until 2000, Novo Nordisk A/S included enzyme producer Novozymes (the two companies "demerged" in 2000); see Novo Nordisk, "Demerger Document." Danisco A/S, a producer of food products, owns enzyme producer Genencor.

[7] These calculations exclude individual inventors and count only corporations, educational institutions, and governments.

[8] For purposes of this study, large companies are those with revenue above $3 billion; medium-sized companies are those with revenue in the range of $100 million to $3 billion; and small companies are those with revenue below $100 million. Company revenue data was obtained from the Orbis Companies Database and corporate Web sites.

[9] The index is constructed by taking the sum of the squares of each company's share in the market. The HHI approaches zero when a market consists of a large number of firms of relatively equal size and increases as the number of firms in the market decreases and as the disparity in their shares increases. Ranges for the HHI depend on how the share is calculated; here, a range of 0 to 1 was used. An index below 0.1 indicates the market

is unconcentrated, an index between 0.1 and 0.18 indicates moderate concentration, and an index above 0.18 indicates the market is highly concentrated. U.S. Department of Justice, "The Herfindahl-Hirschman Index," undated (accessed June 26, 2008).

[10] USITC, Industrial Biotechnology, 2008, 3-28.

[11] Bessen and Meurer, for example, find that small entities realize substantially less value for their patents than large ones. They posit that small firms do not have the resources that large firms have for appropriating value from new technologies and that markets for patents and technology transfer often do not function well thus making it difficult for small firms to license or transfer their technologies to larger ones. Bessen and Meurer, Patent Failure, 2008, 165–86.

[12] Chap. 5 provides suggestions for further research.

[13] Lambiris, Novozymes, "The Importance of Patents," February 11, 2007.

[14] Novozymes Company Web site. http://www.novozymes.com/ (accessed August 29, 2008).

[15] Novozymes, "The Novozymes History," undated (accessed August 28, 2008).

[16] Novozymes, "Novozymes Facts," July 2007.

[17] Ibid.

[18] The exchange rate of $1 = 4.9 120 krone from March 31, 2008, was used to convert currencies. U.S. Department of Treasury Web site. http://www.fms.treas.gov/intn.html#rates (accessed August 29, 2008).

[19] Novozymes, "Novozymes Facts," July 2007.

[20] Novozymes, "Novozymes and Biomass," undated (accessed September 15, 2008).

[21] ICIS Chemical Business, "CRAC Gives Biomass Fuel the Thumbs-Up," July 3, 2006.

[22] Patents and applications are accounted for based on the filing date of the application. With a few minor exceptions, applications that have not resulted in a granted patent do not appear in the data until 2001, when the USPTO first began publishing applications. The numbers presented from the USPTO online database are estimates because they do not include, for example, filings by related companies, filings that contain inventor

information only and cannot be attributed to an assignee, and patents or applications that have been licensed or transferred to or from other companies.

[23] Classification 435/69.1 was one of the classifications with the largest number of patents in the data set provided by the USPTO; however, because it is also a common classification for pharmaceutical, or "red," biotechnology, it is not profiled here.

[24] Daily, "Novozymes and Genencor Settle Dispute," April 19, 2007.

[25] Lambiris, Novozymes, "The Importance of Patents," February 11, 2007.

[26] Verenium Corp, "Form 10-K," March 17, 2008, 2.

[27] Ibid., 11.

[28] Ibid., 2.

[29] Ibid., 36.

[30] Verenium Corp., "Verenium Reports Financial Results for the Second Quarter 2008," August 7, 2008.

[31] Verenium Corp., "Form 10-K," March 17, 2008, 19.

[32] Ibid., 11.

[33] St. Petersburg Times, "Fuel Breakthrough Still Sputters," December 5, 2005.

[34] Verenium Corp., "Form 10-K," March 17, 2008, 3.

[35] Stewart, "Biopolymers," June 2007.

[36] Liu, "Bioplastics in Food Packaging," February 2006, 7.

[37] Schill, "Building Better Bioplastics," June 2007.

[38] In the USPC classification hierarchy, this subclass falls below that of "Micro-organism, tissue cell culture or enzyme using process to synthesize a desired chemical compound or composition (41)," and "Preparing oxygen-containing organic compound (132)." USPTO, "Main Classification Menu," undated (accessed August 11, 2008).

[39] Metabolix, Inc., "Annual Report Form 10-K," 2008, 8.

[40] As discussed infra., the HHI calculation does not account for changes in patent ownership after patents are issued, an important caveat as a substantial number of patents were transferred during the period to Metabolix from another leading patent owner, Monsanto. See Chap. 5 for a discussion of issues for future research.

[41] Metabolix, Inc., "Annual Report Form 10-K," 2008, 3.
[42] Ibid., 8.
[43] Metabolix, Inc., "Quarterly Report Form 10-Q," August 11, 2008, 9.
[44] Mirel Company Web site. http://www.mirelplastics.com/discover/index.html (accessed July 15, 2008).
[45] Metabolix, Inc., "Annual Report Form 10-K," 2008, 18, 20.
[46] Metabolix, "Quarterly Report Form 10-Q." August 11, 2008, 15–16.
[47] Metabolix, Inc., "Annual Report Form 10-K," 2008, 18.
[48] Ibid., 2 1–22.
[49] Ibid., 22.
[50] Ahmann and Dorgan, Bioengineering for Pollution Prevention, 2007, 84.
[51] Metabolix, Inc., "Annual Report Form 10-K," 2008, 39.
[52] Ibid., 8.
[53] In the classification hierarchy, this subclass falls under 41: "Microorganism, tissue cell culture or enzyme using process to synthesize a desired chemical compound or composition"; 132: "Preparing oxygen-containing organic compound"; and 136: "Containing a carboxyl group." USPTO, "Main Classification Menu," undated (accessed August 11, 2008).
[54] See discussion of the applicability of the HHI calculation to patent control infra.
[55] USDA, U.S. Biobased Products, 2008, 100; ETEPS NET, "Consequences, Opportunities and Challenges of Modern Biotechnology," 2007, 41.
[56] Because Cargill is a privately held company, detailed financial data like that provided in the other company profiles are not publicly available.
[57] Datamonitor, "Cargill, Incorporated," November 19, 2007; Cargill, Incorporated Web site, http://www.cargill.com/about/financial/financialhighlights.htm (accessed August 2, 2008).
[58] Schill, "Building Better Bioplastics." June 2007.
[59] Ahmann and Dorgan, Bioengineering for Pollution Prevention, 2007, 50.
[60] NatureWorks LLC, "Welcome to the Blair Manufacturing Operations," March 24, 2007.
[61] Cargill Dow LLC, Agriculture, Nutrition, and Forestry Committee testimony, March 29, 2001.

[62] ATP, "Improving Biodgradable Plastics," undated (accessed August 6, 2008).

[63] Businesswire, "New Corn to Plastic Technology," March 14, 2001.

[64] Cargill Dow LLC, Agriculture, Nutrition, and Forestry Committee testimony, March 29, 2001.

[65] Foundational discoveries related to PLA were patented prior to the 1997–2007 period and thus do not appear in the totals reported in Table 4.6. Moreover, as with the other USPTO classifications profiled, all discoveries in a technology area are not exclusively classified to a particular classification. See chapter 5 for suggestions for further research.

[66] For example see Cargill, "Continuous process for manufacture of lactide polymers with controlled optical purity," U.S. Patent 5,247,058 (September 21, 1993); Cargill, "Melt-stable semi- crystalline lactide polymer film and process for manufacture thereof," U.S. Patent 6,093,791 (July 25, 2000); Cargill, "Degradable polymer fibers; preparation product; and, methods of use," U.S. Patent 6,506,873 (January 14, 2003); and Cargill, "Paper having a melt-stable lactide polymer coating and process for manufacture thereof," U.S. Patent 6,197,380 (March 6, 2001). U.S. Patents and Pre-grant Publications may be accessed through the USPTO online database.

[67] USDA, U.S. Biobased Products, 2008, 77. See also Cargill, "Low pH lactic acid fermentation," U.S. Pre-grant Publication US20030129715 (July 10, 2003); and Cargill, "Low pH lactic acid fermentation," U.S. Pre-grant Publication US20060094093 (October 14, 2005).

[68] For example see NatureWorks, "Fermentation process using specific oxygen uptake rates as a process control," U.S. Patent 7,232,664 (June 19, 2007); NatureWorks, "Methods and materials for the production organic products in cells of Candida species," U.S. Patent 7,141,410 (November 28, 2006); NatureWorks, "Injection stretch blow molding process using polylactide resins," U.S. Pre-grant Publication 20070187876 (December 15, 2006).

[69] USDA, U.S. Biobased Products, 2008, 100; ETEPS NET, "Consequences, Opportunities and Challenges of Modern Biotechnology," 2007, 41.

Chapter 5

CONCLUSIONS

This study relies on patent data, questionnaire results, and technology and firm level data from emerging sectors of industrial biotechnology to provide a detailed picture of innovation in the field. A diverse group of firms, large and small in size, is developing new patented products and processes in industrial biotechnology. The average number of patents held by firms in particular technology areas is small; unlike in other technology areas, leading patenting companies do not hold enormous patent portfolios. Moreover, a steady stream of new firms are patenting each year, suggesting that the field is open to new players, especially the small firms and start-ups that historically have been associated with biotechnology innovation. Moreover, when surveyed by the Commission, firms reported that "patent barriers" were one of the least significant impediments to industrial biotechnology R&D and commercialization efforts.

Profiles of four firms active in particular emerging sectors of industrial biotechnology support these conclusions and further suggest that collaborations—between firms, universities, and government—have become increasingly important in terms of funding, moving technologies from R&D to commercialization, and creating synergies from a combination of different strengths. Patents facilitate these collaborations.

Small firms, like Verenium and Metabolix, hold strong patent portfolios that arise from foundational discoveries made in the university setting and supported by government funding. Bayh-Dole has assisted in this transfer of technologies by permitting universities and firms to retain ownership and license their government-funded discoveries. Verenium and Metabolix also rely on their patent portfolios to attract large firm partners with the capital necessary to develop and commercialize early stage discoveries. Intellectual property is carefully protected in these alliances. Patent portfolios also play an important role in larger firms, like Novozymes and Cargill, which rely on their patents to protect their substantial R&D investments and market share. Thus, patents are playing a fundamental role in the development of the industrial biotechnology field, much as they have in the larger biotechnology sector.

Whether the patenting environment will become more burdensome over time is hard to predict. This study does not consider the impact of the USPTO's substantial backlog of pending applications on firms seeking to patent in the field, nor does it account for the fact that the field of industrial biotechnology is in relatively early stages of development. As more products and processes are commercialized, it is inevitable that the patenting space will become more crowded and contentious. The patent litigation already seen in some leading firms may be a harbinger of more challenging times ahead.

SUGGESTIONS FOR FUTURE RESEARCH

This study has relied on patent data, supplemented with qualitative information, to illustrate the ongoing relationship between patents and innovation in industrial biotechnology. Patent data, however, are imperfect measures of innovation. Many inventions are never patented and many patents are never exploited. Researchers have used various methods to identify those patents that are good measures of innovation and thereby reduce the "noise" inherent in patent data.

One method is to use patent renewal data to identify those patents that have been abandoned for failure to pay renewal fees. Presumably if the owner does not consider the patent worth the fee, it is not a particularly important one. The USPTO requires that patent owners pay fees that increase over the life of the patents at various intervals (3.5, 7.5, and 11.5 years).[1] Bessen found that 60 percent of U.S. patents granted in 1991 were not renewed for the full term because of failure to pay fees; small firm patents were particularly likely not to be renewed.[2] The authors of this study did not obtain renewal data for all patents identified; consideration of these data would enhance the results.[3]

Researchers also have considered the question of whether valuable patents can be identified on their face based on the number of citations to other works and claims that they contain, with more citations and claims suggesting greater value. Some assert that these factors can be used to identify valuable patents based on the patent documents; others are skeptical.[4] The question has not been addressed in this study but merits further attention.

Industrial biotechnology patenting also could be further explored by taking into account patent assignments, transfers, and licenses to obtain a clearer picture of the actual control of patents rights. Patent assignment data is reported by the USPTO; however, licensing information is more difficult to acquire and may be treated as proprietary information by the involved firms. The Metabolix profile demonstrates that the transfer of patent rights may substantially impact a firm's control of a technology area; this issue warrants attention in future research.

Another important caveat to this study was the recognition that industrial biotechnology is an extremely difficult field to identify because of substantial overlap with other sectors and because distinctions drawn by the industry are not reflected in the patent classification system. This study relied on classifications identified in model patents and publicly available data from the USPTO. Proprietary search engines, however, may enable key word searching and other techniques that would yield a more focused picture of industrial biotechnology.

The USPTO will also soon have better searching capabilities available for use with published patent applications; the classification-based search structured for this analysis could be carried out for grant data only and not published application data. Because of the substantial time lag required to obtain a patent grant, improved capabilities for more fully accessing published application data would be extremely useful to researchers and, more importantly, to firms that are attempting to understand their operational freedom in the face of substantial patenting activity in industrial biotechnology.

ENDNOTES

[1] USPTO, "FY 2009 Patent and Fee Schedule," October 1, 2008.

[2] Bessen, "The Value of U.S. Patents," 2006, 27; Bessen and Meurer, Patent Failure, 2008, 100.

[3] Many of the patents identified are relatively new and thus have only come up for renewal once or twice. The authors did not locate any nonrenewed patents in spot checks; however, renewal data were not consulted for all identified patents.

[4] For more information, compare Allison, et al., "Valuable Patents," 2004, 435, and Allison and Sager, "Valuable Patents Redux," 2007, 1769, with Adelman and DeAngelis, "Patent Metrics," 2007, 1716–29.

BIBLIOGRAPHY

Adelman, David E., and Kathryn L. DeAngelis. "Patent Metrics: The Mismeasure of Innovation in the Biotech Patent Debate." Texas Law Review 85 (2007): 1677–1744.

Advanced Technology Program (ATP). "Improving Biodgradable Plastics Manufactured from Corn." ATP Status Report Advanced Materials and Chemicals 94-01-0173, undated. http://statusreports.atp.nist.gov/reports/94-01-0173PDF.pdf (accessed August 6, 2008).

Ahmann, Dianne, and John R. Dorgan. Bioengineering for Pollution Prevention through Development of Biobased Energy and Materials: State of the Science Report. Washington, DC: EPA, 2007. http://es.epa.gov/ncer/publications/statesci/bioengineering.pdf.

Allison, John R., Mark A. Lemley, Kimberly A. Moore, and Derek R. Trunkey. "Valuable Patents." Georgetown Law Journal 92 (2004): 435–75.

Allison, John R., and Thomas W. Sager. "Valuable Patents Redux: On the Enduring Merit of Using Patent Characteristics to Identify Valuable Patents." Texas Law Review 85 (2007): 1769–97.

Association of University Technology Managers (AUTM). A UTM U.S. Licensing Activity Survey: FY 2006, 2007. http://www.autm.net/events/file/AUTM _06 _US%20LSS _FNL.pdf.

———. AUTM U.S. Licensing Survey: FY 1991–FY 1995, 1996. http://www.autm.net/events/File/Surveys/91 - 95AUTMLicSurveyPublic.pdf.

Barfield, Claude, and John E. Calfee. Biotechnology and the Patent System. Washington, DC: AEI Press, 2007.

Bessen, James E. "The Value of U.S. Patents by Owner and Patent Characteristics." Boston University School of Law Working Paper Series. Law and Economics, Working Paper no. 06-46, 2006.

Bessen, James. E., and Michael J. Meurer. Patent Failure. Princeton, NJ: Princeton University Press, 2008.

Biotechnology Industry Organization (BIO). "Biotechnology: A Collection of Technologies." Guide to Biotechnology 2008, 2008. http://www.bio.org/speeches/pubs/er/technology collection.asp.

———. "Industrial and Environmental Applications." Guide to Biotechnology 2008, 2008. http://bio.org/speeches/pubs/er/industrial.asp.

———. "The Third Wave in Biotechnology," 2008. http://bio.org/ind/background/thirdwave.asp.

Block, Fred, and Matthew R. Keller. Where Do Innovations Come From? Transformations in the U.S. National Innovation System, 19 70-2006. The Information Technology & Innovation Foundation, July 2008. http://www.itif.org/files/Where_do_innovations_come_from.pdf.

Bureau van Dijk. Orbis Companies Database. https://orbis.bvdep.com (accessed June– September 2008).

Businesswire. "New Corn to Plastic Technology Receives Department of Energy Honors," March 14, 2001.

http://findarticles.com/p/articles/mi m0EIN/is 2001 March 14/ai 71707215.

Daily, Robert. "Novozymes and Genencor Settle Dispute Over Alternative Fuel Technology." Patent Docs, April 19, 2007. http://www.patentdocs.net/patent_docs/2007/04/index.html.

Datamonitor. "Cargill, Incorporated." November 19, 2007. http://www.datamonitor.com.

European Techno-Economic Policy Support Network (ETEPS NET). "Consequences, Opportunities and Challenges of Modern Biotechnology for Europe (BIO4EU) – Task 2." Annex to Report 3, Deliverable 21, 2007. http://bio4eu.jrc.ec.europa.eu/documents/Bio4EUTask2Annexindustrialproduction.pdf.

Federal Trade Commission (FTC). To Promote Innovation: The Proper Balance of Competition and Patent Law Policy. Washington, DC: FTC, 2003.

Fromer, Jeanne C. "Patent Disclosure." Iowa Law Review 94 __ (2009) (forthcoming). http://papers.ssrn.com/sol3/papers.cfm?abstract id= 1116020.

Gruber, Patrick R., Cargill Dow LLC. Testimony before the U.S. Senate Committee on Agriculture, Nutrition and Forestry, March 29, 2001. http://agriculture.senate.gov/Hearings/Hearings 2001/March 29 200 1/0329gru.htm.

Hahn, Robert W. "An Overview of the Economics of Intellectual Property Protection." In Intellectual Property Rights in Frontier Industries: Software and Biotechnology, edited by Robert W. Hahn, 11–44. Washington, D.C.: AEI Brookings Joint Center for Regulatory Studies, 2005.

Heller, Michael, and Rebecca Eisenberg. "Can Patents Deter Innovation? The Anticommons in Biomedical Research." Science 280 (May 1, 1998): 698–701.

Henderson, Rebecca, Adam B. Jaffe, and Manuel Trajtenberg. "Universities as a Source of Commercial Technology: A Detailed Analysis of University Patenting, 1965-1988." The Review of Economics and Statistics, no. 1 (February 1998): 119–27.

ICIS Chemical Business. "CRAC Gives Biomass Fuel the Thumbs-up." Vol. 1, issue 26 (July 3, 2006): 13.

Jaffe, Adam B., and Josh Lerner. Innovation and Its Discontents: How Our Broken Patent System Is Endangering Innovation and Progress, and What to Do About It. Princeton, NJ: Princeton University Press, 2004.

———. "Reinventing Public R&D: Patent Policy and the Commercialization of National Laboratory Technologies." The Rand Journal of Economics 32, no. 1 (2001): 167–98.

Jenson, Richard, and Marie Thursby. "Proofs and Prototypes for Sale: The Licensing of University Inventions." American Economic Review 91, no. 1 (March 2001): 240–59.

Kirk, Ole, Ture Damhus, Torben Vedel Borchert, Claus Crone Fuglsang, Hans Sejr Olsen, Tomas Tage Hansen, Henrik Lund, Hans Erik Schiff, and Lone Kierstein Nielsen. "Enzyme Applications, Industrial." In Kirk-Othmer Encyclopedia of Chemical Technology, vol. 10, 248–317. Hoboken, NJ: John Wiley & Son, 2004.

Knauf, Michael, and Mohammed Moniruzzaman. "Lignocellulosic Biomass Processing: A Perspective." International Sugar Journal 106, no. 1263 (2004): 147–50.

Lambiris, Elias J. Novozymes. "The Importance of Patents," February 11, 2007. http://www.novozymes.com/NR/rdonlyres/9154B681-ACBB-4660-BE4A46514340 19A7/0/5TheimportanceofpatentsCMD2007 FINAL.pdf.

Lerner, Joshua. "The Importance of Patent Scope: an Empirical Analysis." RAND Journal of Economics 25, no. 2 (1994): 319–33.

———. "Small Businesses, Innovation, and Public Policy." In Are Small Firms Important? edited by Zolton Acs, 159-68. New York: Kluwer, 1999.

Liu, Lillian. "Bioplastics in Food Packaging: Innovative Technologies for Biodegradable Packaging," February 2006.

Merritt, Rick. "Fixing the Patent Office." EE Times, September 17, 2007. http://www.eetimes.com/showArticle.jhtml?articleID=201806944.

Metabolix, Inc. "Annual Report Form 10-K," 2008. http://files.shareholder.com/downloads/MBLX/359622896x0x220896/A0A489F3-65DC4C 1F-BB 1 0-77E7CC9DFDF3/metabolix.pdf.

———. "Quarterly Report Form 10-Q," August 11, 2008. http://ir.metabolix.com/secfiling.cfm?filingID=1104659-08-51879.

National Academy of Sciences (NAS). Industrial Research and Innovation Indicators: Report of a Workshop. Washington, DC: NAS, 1997.

National Science Board (NSB). Science and Engineering Indicators 2008. Arlington, VA: NSF, 2008.

———. Science and Engineering Indicators 1993. Arlington, VA: NSF, 1993.

Nature Works LLC. "Welcome to the Blair Manufacturing Operations of NatureWorks LLC," March 24, 2007.

Novo Nordisk. "Demerger Document," undated. http://wwwprod.novonordisk.com/pdf/Press_News_English_Attachments/001016_01_Demer ger.pdf. (accessed September 12, 2008).

Novozymes. "Novozymes and Biomass," undated. http://biomass.novozymes.com/faq/ (accessed September 15, 2008).

———. "Novozymes Facts," July 2007. http://www.novozymes.com/NR/rdonlyres/E71908D9- 3D00-455B-9 1 9D-A2 1F3C436 1 6C/0/Nzs UK Tryk web.pdf.

———. "The Novozymes History," undated. http://www.novozymes.com/en/MainStructure/AboutUs/Our+history/ (accessed August 28, 2008).

Schacht, Wendy. R&D Partnerships and Intellectual Property: Implications for U.S. Policy. CRS Report for Congress, 2000.

Schill, Susanne Retka. "Building Better Bioplastics." Biomass Magazine, June 2007. http://www.biomassmagazine.com/article.jsp?article id=1158.

Shapiro, Carl. "Navigating the Patent Thicket: Cross Licenses, Patent Pools, and Standard- Setting." In Innovation Policy and the Economy, edited by Adam B. Jaffee, Josh Lerner, and Scott Stern, 119–50. Cambridge, MA: NBER, 2001.

Stewart, Richard. "Biopolymers." Plastics Engineering, June 2007.

St. Petersburg Times. "Fuel Breakthrough Still Sputters," December 5, 2005. http://www.sptimes.com/2005/12/05/State/Fuel breakthrough sti.shtml.

U.S. Department of Agriculture (USDA). U.S. Biobased Products Market Potential and Projections through 2025. Washington, DC: USDA, 2008.

U.S. Department of Justice (U.S. DOJ). "The Herfindahl-Hirschman Index," undated. http://www.usdoj.gov/atr/public/testimony/hhi.htm (accessed June 26, 2008).

United States Government Accountability Office (U.S. GAO). U.S. Patent and Treadmark Office: Hiring Efforts Are Not Sufficient to Reduce the Patent Application Backlog. GAO-07-1102, September 2007. http://www.gao.gov/new.items/d071102.pdf.

U.S. International Trade Commission (USITC). Industrial Biotechnology: Development and Adoption by the U.S. Chemical and Biofuel Industries. USITC Publication 4020. Washington, DC: USITC: 2008.

U.S. Patent and Trademark Office (USPTO). "FY 2009 Patent and Fee Schedule."
October 1, 2008. http://www.uspto.gov/web/offices/ac/qs/ope/fee2008 october02.htm#top.
———. "Main Classification Menu," undated. http://www.uspto.gov/web/patents/classification/ (accessed August 11, 2008).
———. Overview of the U.S. Patent Classification System (USPC). June 2008. http://www.uspto.gov/web/offices/opc/documents/overview.pdf.
———. Patent Technology Center Groups 1630–1660, Biotechnology: All Classified Patents. Washington, DC: USPTO, 2006 (on file with authors).
———. Performance and Accountability Report Fiscal Year 2007, December 21, 2007. http://www.uspto.gov/web/offices/com/annual/2007/index.html.
_____. USPTO Patent Full-Text and Image Database. http://patft.uspto.gov/ (accessed June– September 2008).
_____. Utility Patents in USITC-Identified Subclasses Based on Primary Patent Classification, Washington, DC: USPTO, 2007 (on file with authors).
———. What Are Patents, Trademarks, Servicemarks, and Copyrights? May 12, 2004. http://www.uspto.gov/web/offices/pac/doc/general/whatis.htm.
Verenium Corp., "Form 10-K," March 17, 2008. http://media.corporate-ir.net/media files/irol/8 1/81 345/annualreports/2007 1 0K.pdf.
_____. "Next Generation Cellulosic Ethanol." undated http://www.verenium.com/pdf/Veren_NextGenCellulosicEthanol.pdf (accessed October 1, 2008).
———. "Verenium Reports Financial Results for the Second Quarter 2008," August 7, 2008. http://ir.verenium.com/phoenix.zhtml?c=81345&p=irol-newsArticle&ID= 118483 7&highlight=.
Walsh, John P., Ashish Arora, and Wesley M. Cohen. "Effects of Research Tools Patents and Licensing on Biomedical Innovation." In Patents in the Knowledge-Based Economy, edited by Wesley M. Cohen and Stephen A. Merrill, 285–336. Washington, DC: National Academies Press, 2003.
Walsh, John P., Charlene Cho, and Wesley M. Cohen. "View from the Bench: Patents and Material Transfers." Science 309 (September 23, 2005): 2002–03.

Wolters Kluwer Health. "IFI Patent Intelligence Announces 2007's Top U.S. Patent Assignees." News release, January 14, 2008. http://www.ificlaims.com/IFI%20Patent%20Release%201-9-08.htm.

APPENDIX A.
PATENT SEARCH METHODOLOGY

The custom data set provided by the USPTO consists of 20,418 patents granted during the 1975–2006 period in USPC classifications identified by the authors as relevant to industrial biotechnology. The authors employed the following methodology to obtain a list of USPC classifications relevant to industrial biotechnology. First, and based on research conducted in connection with the Commission's report on industrial biotechnology, model industrial biotechnology patents were identified in the following areas: biofuels, biopolymers, chemical processes, micro-organisms, enzymes, and pharmaceuticals. All primary and secondary classifications appearing in these model patents were also identified. The USPC classifications are technology and science based rather than industry based, and thus responsive search results included patents for the identified technologies regardless of the industry where they were applied.

Next, the USPTO's online database was used to test these classifications and identify those that seemed to produce the most patents relevant to industrial biotechnology using random sampling techniques. In addition to the classifications identified in this manner, and to account for the fact that the USPC is a hierarchical system, the list of classifications was expanded to include all "children," i.e., those that fell below the identified classifications in the hierarchy. The USPTO then ran the requested search of all patents with a primary classification equivalent to one of the identified classifications.

The identified classifications were the following:

127/037	435/070.500	435/101	435/144	435/205	435/254.600	
435/041	435/071.100	435/102	435/145	435/206	435/254.700	
435/042	435/071.200	435/103	435/146	435/207	435/254.900	
435/043	435/071.300	435/104	435/147	435/208	435/255.100	
435/044	435/072	435/105	435/148	435/209	435/255.200	
435/045	435/073	435/105	435/149	435/2 10	435/255.210	
435/046	435/074	435/105	435/150	435/211	435/255.300	
435/047	435/075	435/106	435/155	435/2 12	435/255.400	
435/048	435/076	435/107	435/156	435/2 13	435/255.500	
435/049	435/077	435/108	435/157	435/214	435/255.600	
435/050	435/078	435/109	435/158	435/2 15	435/255.700	
435/051	435/079	435/110	435/159	435/2 16	435/256.100	
435/052	435/080	435/111	435/160	435/2 17	435/256.300	
435/053	435/081	435/112	435/161	435/2 18	435/256.500	
435/054	435/082	435/113	435/162	435/2 19	435/256.600	
435/055	435/083	435/114	435/163	435/220	435/256.700	
435/056	435/084	435/115	435/165	435/221	435/256.800	
435/057	435/085	435/116	435/166	435/222	435/280	
435/058	435/086	435/117	435/167	435/223	435/320.100	
435/059	435/087	435/118	435/168	435/224	435/410	
435/060	435/088	435/119	435/169	435/225	435/411	
435/061	435/089	435/1 20	435/170	435/226	435/412	
435/062	435/090	435/121	435/171	435/227	435/414	
435/063	435/091.100	435/1 22	435/183	435/228	435/415	
435/064	435/091.200	435/1 23	435/184	435/229	435/417	
435/065	435/091.210	435/124	435/185	435/230	435/418	
435/066	435/091.300	435/1 25	435/186	435/231	435/419	
435/067	435/091.310	435/1 26	435/187	435/232	435/420	
435/068.100	435/091.320	435/1 27	435/188	435/233	435/42 1	
435/069.100	435/091.330	435/1 28	435/188.500	435/234	435/422	

Appendix A

435/069.200	435/091.400	435/1 29	435/189	435/252.300	435/423
435/069.300	435/091.410	435/130	435/190	435/252.310	435/424
435/069.400	435/091.500	435/131	435/191	435/252.320	435/425
435/069.500	435/091.510	435/132	435/192	435/252.330	435/426
435/069.510	435/091.520	435/133	435/193	435/252.340	435/427
435/069.520	435/091.530	435/134	435/194	435/252.350	435/428
435/069.600	435/092	435/135	435/195	435/254.100	435/429
435/069.700	435/093	435/136	435/196	435/254.110	435/430
435/069.800	435/094	435/137	435/197	435/254.200	435/430.100
435/069.900	435/095	435/138	435/198	435/254.210	435/431
435/070.100	435/096	435/139	435/199	435/254.220	536/023.200
435/070.200	435/097	435/140	435/200	435/254.230	536/023.700
435/070.210	435/098	435/141	435/201	435/254.300	536/023.710
435/070.300	435/099	435/142	435/203	435/254.400	536/023.720
435/070.400	435/100	435/143	435/204	435/254.500	536/023.740

The search results are reported in USPTO, Utility Patents in USITC-Identified Subclasses Based on Primary Patent Classification, 2007. Access to the search results are available via the following link: Utility Patents In USITC Identified Subclasses.xls.

INDEX

A

access, 55
acid, 9, 43, 44, 50, 54, 59
acquisitions, 31
AEI, 66, 67
agricultural, 22, 31, 47, 48, 50, 53
agriculture, 8, 67
alpha, 40, 50
alternative, 41
alternatives, 47
amylase, 40
application, 23, 39, 43, 49, 54, 55, 56, 63
assignment, 63
assumptions, 37
ATP, 59, 65
Australia, 38

B

bacteria, 12, 42
bacterial, 44, 47
bacterial fermentation, 44
bacterium, 14
barriers, vii, 3, 7, 8, 11, 14, 21, 29, 30, 61
benchmark, 16
benefits, 43
beverages, 9
biocatalysts, vii, 1, 8, 9
bioengineering, 65
biofuel, vii, 2, 3, 7, 8, 10, 11, 19, 21
biofuels, 1, 2, 4, 8, 9, 10, 12, 30, 33, 38, 41, 47, 73
biomass, 2, 8, 33, 34, 41, 47, 48, 69
biomolecular, 8
biopolymers, 30, 73
Biopolymers, 57, 69
biosynthesis, 47, 50
biotechnology, vii, viii, 1, 2, 3, 4, 5, 7, 8, 9, 10, 11, 12, 13, 14, 15, 16, 17, 18, 21, 22, 23, 24, 25, 26, 27, 28, 29, 30, 33, 34, 37, 38, 40, 43, 47, 54, 57, 61, 62, 63, 64, 73
blocks, 17
Boston, 66
Brazil, 38
breakdown, 42
British Columbia, 36
building blocks, 44

C

Canada, 35, 36

Candida, 59
capacity, viii, 2, 53
carbohydrates, 22
carboxyl, 58
Carboxylic acid, 44
cash flow, 41, 48
cell, 57, 58
cell culture, 57, 58
cellulose, 34, 38, 42
cellulosic, viii, 2, 3, 4, 8, 9, 22, 33, 34, 38, 41, 42
cellulosic ethanol, viii, 3, 4, 8, 9, 22, 34, 38, 41, 42
chemical industry, 9
chemical reactions, 8, 9, 12
chemicals, 1, 2, 8, 9, 11, 12, 33, 47, 48
children, 30, 73
China, 38
classes, 15, 22, 30, 40, 53
classification, vii, 2, 3, 4, 5, 10, 11, 17, 21, 22, 30, 34, 35, 36, 37, 40, 44, 45, 46, 50, 51, 52, 55, 57, 58, 59, 63, 70, 74
Co, 52
coatings, 49
commercialization, vii, viii, 1, 3, 4, 5, 7, 8, 11, 13, 16, 19, 21, 29, 34, 40, 42, 47, 48, 52, 53, 61
commons, 18, 20
competition, 14
competitive conditions, 29
competitor, 4
composition, 57, 58
compounds, 22
concentration, 56
confidentiality, 49
Congress, iv, 14, 15, 69
consumers, 43
contractors, 14
contracts, 42
control, viii, 4, 36, 37, 46, 51, 58, 63
copolymer, 50
corn, 9, 34, 53

corporations, 5, 25, 26, 31, 33, 34, 44, 46, 50, 55
cost-effective, 3, 34, 54
costs, 1, 20
country of origin, 35
covering, 48, 54
crops, 9, 10, 48
CRS, 69
crystalline, 59
culture, 54

D

data set, 7, 10, 57, 73
database, 10, 18, 21, 33, 42, 49, 55, 56, 59, 73
dating, 53
Degussa, 28
demand, 1
Denmark, 28, 36, 38, 46, 51, 52
Department of Agriculture, 48, 69
Department of Commerce, 48
Department of Energy, 38, 66
Department of Energy (DOE), 38
Department of Justice, 56, 69
detergents, 9
Diamond, 14
disclosure, 13
disputes, 40
distillation, 50
diversity, 29, 54
DNA, 11, 40, 43
drug delivery, 50
drug delivery systems, 50
DSM, 36

E

E. coli, 12, 42
eating, 14
economic performance, 9

Index

educational institutions, 34, 44, 55
employees, 38, 53
energy, 1, 48
environment, 18, 62
environmental impact, 8
Environmental Protection Agency, 50
enzymatic, 2, 8, 10
enzymes, vii, viii, 1, 2, 3, 4, 8, 9, 11, 12, 22, 29, 33, 34, 38, 40, 41, 43, 55, 73
EPA, 65
ester, 44
ethanol, viii, 3, 4, 8, 9, 22, 34, 38, 40, 41, 42, 47
Ethanol, 40, 55, 70
Europe, 51, 66
evolution, 43
exchange rate, 56
expenditures, 38, 39, 41, 48
expert, iv

F

failure, 62
family, 52
February, 19, 56, 57, 67, 68
federal funds, 16
federal government, viii, 4, 5, 14
Federal Trade Commission, 14, 67
Federal Trade Commission (FTC), 14, 67
fee, 62
feedstock, 32
fees, 62
fermentation, 9, 42, 44, 47, 50, 54, 59
fibers, 50, 59
film, 44
films, 49
financing, 13, 16
firm size, 29, 32
firms, vii, viii, 1, 2, 4, 5, 7, 8, 10, 11, 14, 16, 18, 19, 21, 25, 27, 29, 30, 36, 38, 40, 55, 56, 61, 62, 63, 64

flow, 41, 48
focusing, 21
food, 9, 34, 50, 53, 55
food production, 55
food products, 55
Forestry, 58, 59, 67
France, 51, 52
freedom, 4, 38, 40, 64
FTC, 19, 67
fuel, 29, 47
funding, 5, 14, 42, 48, 52, 53, 61, 62
funds, 38

G

gene, 17, 43
General Agreement on Tariffs and Trade, 23
generation, 43
genes, 47, 49
genetically modified organisms, 34
Georgia, 36
Germany, 28, 35, 44, 50, 51, 52
glycosyl, 55
goals, 48
government, viii, 2, 3, 4, 5, 14, 15, 16, 26, 27, 31, 34, 38, 40, 42, 44, 46, 47, 48, 52, 53, 55, 61, 62
Government Accountability Office, 69
grain, 53
grants, 5, 10, 22, 23, 24, 25, 39, 40, 41, 42, 48
greenhouse, 1
greenhouse gas, 1
groups, 17
growth, 11, 14, 38

H

HA, 44
hands, 3

health, 41
hemicellulose, 34
high tech, 3, 9
hiring, 24
household, 9
hydrolases, 55
hydrolysis, 9
hypothesis, 18

I

IBM, 29
id, 67, 69
identity, 31
implementation, 3
incentive, 20
income, 38, 39
India, 38, 44, 46
industrial, vii, viii, 1, 2, 3, 4, 5, 7, 8, 9, 10, 11, 12, 15, 18, 21, 22, 23, 24, 25, 26, 27, 29, 30, 33, 34, 37, 38, 40, 41, 43, 47, 53, 54, 61, 62, 63, 64, 66, 73
industrial processing, 9
industry, 3, 9, 11, 12, 14, 15, 16, 27, 29, 32, 38, 47, 63, 73
information technology, 3
Information Technology, 66
infringement, 18, 40
injury, iv
innovation, vii, 1, 2, 5, 7, 8, 10, 11, 13, 16, 17, 24, 61, 62
Innovation, i, iii, v, 12, 13, 15, 19, 20, 65, 66, 67, 68, 69, 70
institutions, 34, 44, 55
intellectual property, 16, 17, 30, 31, 40, 42, 48, 49, 50
intellectual property rights, 49
International Trade, vii, 3, 69
International Trade Commission, vii, 3, 69
interpretation, 10, 18, 37
inventions, 3, 10, 14, 15, 16, 62
inventors, 15, 17, 23, 55
investment, 16, 40
investors, 16
isolation, 50

J

January, vii, 2, 18, 21, 22, 23, 30, 32, 50, 59, 71
Japan, 35, 36, 44, 46, 51, 52
Japanese, 53

K

kernel, 34

L

lactic acid, 50, 54, 59
large-scale, 47
law, 14
laws, 49
leather, 9, 38
legislation, 15
legislative, 15
licenses, 15, 17, 42, 49, 63
licensing, 14, 15, 18, 37, 63
life forms, 12, 14
likelihood, 16
limitations, 10
lipids, 44
litigation, 4, 62
location, 44, 51
Lockheed Martin, 36
losses, 41
Louisiana, 41

M

magnetic, iv

management, 53
manipulation, 9
manufacturing, 1
market, viii, 1, 4, 5, 13, 16, 36, 38, 40, 50, 55, 62
market concentration, 36
market position, 5
market share, viii, 4, 5, 38, 40, 62
markets, 54, 56
Massachusetts, 4, 47
Massachusetts Institute of Technology, 4, 47
matching funds, 38
measures, 62
media, 70
median, 28, 35, 44, 51
melt, 59
mergers, 31
metabolic, 50
metabolic pathways, 50
metabolism, 12
microbes, 49
microbial, 43, 47
microorganisms, vii, 1
micro-organisms, 8, 9, 12, 22, 38, 43, 73
Millennium, 28
mirror, viii, 2, 20, 23
MIT, 4, 47, 49, 50
Mitsubishi, 52
molecular biology, 22, 34, 40, 43, 44, 50
movement, 1, 8, 13

N

NAS, 12, 17, 68
National Academy of Sciences, 68
National Academy of Sciences (NAS), 68
National Institute of Standards and Technology, 53
National Institutes of Health, 42
natural, 9

Nebraska, 53
net income, 38
Netherlands, 35, 36, 51, 52
New York, iii, iv, 68
Nielsen, 68
noise, 37, 62
nucleic acid, 43
nutrition, 41

O

online, 10, 33, 42, 47, 49, 55, 56, 59, 73
optical, 59
optimization, 43
organic, 12, 57, 58, 59
organic compounds, 12
organism, 57, 58
organizations, 31
ownership, 3, 8, 21, 31, 37, 57, 62
oxygen, 57, 58, 59

P

packaging, 50
paints, 8
paper, 44
Paper, 59, 66
partnership, 42
Patent and Trademark Office, vii, 2, 70
patents, vii, viii, 1, 2, 3, 4, 5, 7, 8, 10, 11, 13, 14, 15, 16, 17, 18, 21, 22, 23, 24, 25, 26, 27, 28, 29, 30, 31, 33, 34, 35, 36, 37, 39, 40, 42, 43, 44, 46, 49, 50, 51, 52, 53, 54, 56, 57, 61, 62, 63, 64, 70, 73
pathways, 50
performance, 9, 38, 53
petrochemical, 47
petroleum, 1, 14, 43, 53
pH, 54, 59
pharmaceutical, 17, 22, 38, 57

pharmaceuticals, 9, 10, 29, 30, 73
plants, 49
plastic, 5, 50
plastics, viii, 2, 3, 4, 5, 8, 9, 11, 22, 33, 43, 47, 48, 52, 53
play, vii, 2, 4, 5, 7, 8, 43, 62
polyester, 50
polymer, 53, 54, 59
polymer film, 59
polymer synthesis, 54
polymerization, 50
polymers, 49, 50, 59
polypeptide, 11, 40
polysaccharides, 44
portfolio, 4, 37, 47, 52, 54
portfolios, vii, 1, 3, 5, 7, 10, 11, 52, 54, 61, 62
power, 47
prices, 32, 43
process control, 59
producers, 47
production, viii, 2, 3, 4, 8, 9, 10, 22, 33, 34, 38, 40, 41, 42, 47, 48, 50, 53, 54, 59
program, 47, 48
proliferation, 17
property, iv, 16, 17, 20, 30, 31, 40, 42, 48, 49, 50, 62
property rights, 17
protection, 13
protein, 11, 40
prototype, 16
public, 13, 18, 69
purification, 34, 50

Q

questionnaire, 2, 5, 8, 10, 11, 19, 21, 29, 61

R

R&D, vii, viii, 1, 2, 3, 4, 5, 7, 8, 10, 11, 15, 16, 19, 21, 28, 29, 30, 32, 37, 38, 39, 40, 41, 43, 47, 48, 53, 54, 61, 62, 67, 69
R&D investments, viii, 4, 5, 37, 48, 53, 62
random, 73
range, 28, 32, 36, 44, 46, 51, 52, 55
raw material, 12
raw materials, 12
receptors, 17
recognition, 63
recovery, 54
reduction, 18, 38
regulatory requirements, 23, 32
relationship, 7, 13, 31, 62
relationships, 32
renewable resource, 1, 8, 43
research, vii, 1, 4, 12, 13, 14, 15, 16, 17, 31, 37, 38, 48, 56, 57, 59, 63, 73
research and development, vii, 1
researchers, 17, 18, 50, 63
resins, 59
resources, 1, 8, 40, 43, 56
revenue, 28, 36, 38, 46, 52, 53, 55
risk, 32, 53
risk management, 53

S

sales, 31, 38, 40
salt, 50
sample, 16, 43
sampling, 73
Samsung, 29
Schiff, 68
scientists, 17, 18, 47, 49
search, 2, 7, 12, 18, 21, 22, 30, 63, 73, 74, 75
search engine, 63

Index

search terms, 22
searches, 10
searching, 63
secret, 49
security, 34
semiconductor, 29
senate, 67
Senate, 29, 67
separation, 31
series, 17
services, iv, 53, 54
shares, 55
sites, 55
small firms, 14, 29, 37, 56, 61
solutions, 18
solvents, 44
species, 59
spin, 14, 15
sporadic, 46
St. Petersburg, 57, 69
stages, 62
standards, 24
starch, 9, 22, 38
storage, 53
strain, 47
strategic, viii, 2, 4, 5, 38, 40, 41, 47, 52
strength, 52, 54
sugars, 9, 34, 38, 42, 44, 50
Supreme Court, 14
Sweden, 38
Switzerland, 36, 44
synthesis, 54
systems, 47, 49, 50

T

technology, vii, viii, 1, 3, 4, 5, 9, 12, 13, 15, 16, 17, 18, 19, 22, 32, 33, 34, 36, 40, 41, 42, 46, 47, 50, 51, 56, 59, 61, 63, 66, 73
technology transfer, 13, 15, 56

telecommunications, 29
testimony, 58, 59, 69
Texas, 65
textiles, 9, 55
theory, 20
threatening, 18
time, 13, 15, 18, 23, 36, 46, 62, 63
tissue, 22, 57, 58
title, 14, 15
total revenue, 53
trade, 13, 49
trade-off, 13
trading, 53
transaction costs, 20
transactions, 37
transfer, 4, 5, 13, 15, 16, 40, 56, 62, 63
transfer of technologies, 62
transgenic, 47
Treasury, 56
trend, 46

U

U.S. Department of Agriculture, 48, 69
U.S. Department of Agriculture (USDA), 69
UK, 69
United Kingdom, 38, 52
United States, 5, 28, 29, 36, 38, 44, 46, 50, 51, 52, 54, 69
universities, 3, 14, 15, 16, 18, 25, 26, 31, 44, 46, 61, 62
Uruguay, 23
Uruguay Round, 23
USDA, 58, 59, 69

V

validity, 50
vitamin B1, 9
vitamin B12, 9

W

warrants, 63
web, 17, 69, 70
winning, 16

Y

yeast, 12
yield, 63